P9-AZV-315

Acclaim for

ANTHONY SHADID

and

HOUSE OF STONE

"This is not just the Arab world's *Year in Provence.* It is as if Shadid has combined the breakthrough effects of Amy Tan's *The Joy Luck Club,* William Faulkner's *As I Lay Dying,* and Frances Fitzgerald's *Fire in the Lake* into one enormously likeable book."
— **John Freeman,** *Boston Globe*

"Elegiac, heartbreaking. . . . A book conceived as an introspective project of personal recovery — as well as a meditation on politics, identity, craft, and beauty in the Levant — now stands as a memorial. It is a fitting one because of the writing skill and deep feeling [Shadid] unobtrusively displays."   — **Steve Coll,** *New York Times*

"An apt testament — a moving contemplation of how the dead stay with us, and how war scrambles the narrative of family life."
— *The New Yorker*

"Profound, insightful, tragic, and funny. . . . There is not space here to sell out all of this book's many rewards. . . . The prose is ripe, the biblical landscapes vividly rendered."
— *Telegraph* **(London)**

"Both intimate and sweeping. . . . An extraordinary memoir, rooted in humility and humanity."   — *Seattle Times*

"Beautifully written. . . . [Shadid's] book brims with zest for life, for savoring the present moment, for all things handcrafted."
— *Denver Post*

"Shadid seems to have felt a longing to reassert a link to something permanent, to breathe new life into the idea of home. . . . Poignant."
— *New York Review of Books*

"A diary of architectural adventure, a personal record of family history, a subtle examination of intricate regional politics, and an Odyssean journey home. . . . In the subtlest of his delicate metaphors, Shadid remakes the house from the very materials of shattered Lebanon, remnants of war reassembled in a structure of peace."
— *American Prospect*

"Deft, witty, and deeply moving. . . . Shadid writes eloquently, and without sentimentality."   — *Oregonian*

"A fascinating portrait of a unique time and place."
— *Minneapolis Star-Tribune*

"[A] moving memoir. . . . Shadid offers a carefully observed meditation not only on Lebanese village life, but on what it is like to try and build a sense of home in the midst of a war zone."
— *Globe and Mail* (Toronto)

"Ultimately a search for identity disguised as a book about interior design. . . . *House of Stone* is a moving portrait of the human heart, and its resilience against wars and changing empires."
— *Forbes*

"Sober and intelligent. . . . This fine book stands as a fitting and poignant memorial to a remarkable journalist."
— *Financial Times*

"Shadid's great skill as a journalist was that of a master storyteller, and he's never been more effective than in his final book."
— *Bookforum*

"An honest-to-god, hands-down, undeniable and instant classic . . . written with levity and candor and lyricism that makes the book, improbably, both a compulsive read and one you don't want to end."
— Dave Eggers, author of *A Hologram for the King*

"Poignant, aching, and at times laugh-out-loud funny. . . . Shadid's writing is so lyrical it's like hearing a song."
— David Finkel, author of *The Good Soldiers*

"In rebuilding his family home in southern Lebanon, Shadid commits an extraordinarily generous act of restoration for his wounded land, and for us all."— Annia Ciezadlo, author of *Days of Honey*

"Evocative and beautifully written, House of Stone . . . should be read by anyone who wishes to understand the agonies and hopes of the Middle East."— Kai Bird, author of *Crossing Mandelbaum Gate*

"[A] beautifully rendered memoir. . . . The rooms and hallways of [Shadid's] great-grandfather's house tell stories that will linger with every reader for decades."— André Aciman, author of *Out of Egypt*

Books by Anthony Shadid

Legacy of the Prophet: Despots, Democrats,
and the New Politics of Islam

Night Draws Near: Iraq's People in the
Shadow of America's War

House of Stone: A Memoir of Home,
Family, and a Lost Middle East

HOUSE
OF
STONE

A MEMOIR OF HOME, FAMILY,
AND A LOST MIDDLE EAST

Anthony Shadid

WITHDRAWN

Jefferson Madison
Regional Library
Charlottesville, Virginia

Mariner Books
Houghton Mifflin Harcourt
BOSTON • NEW YORK

First Mariner Books edition 2013

Copyright © 2012 by Anthony Shadid

All rights reserved

For information about permission to reproduce selections from this
book, write to Permissions, Houghton Mifflin Harcourt Publishing
Company, 215 Park Avenue South, New York, New York 10003.

www.hmhbooks.com

Library of Congress Cataloging-in-Publication Data
Shadid, Anthony.
House of stone: a memoir of home, family, and a lost Middle East / Anthony Shadid.
p. cm.
ISBN 978-0-547-13466-6 (hardback)   ISBN 978-0-544-00219-7 (pbk.)
1. Families — Lebanon. 2. Home — Lebanon — History. 3. Lebanon — Emigration
and immigration — Social aspects. 4. Middle East — Social conditions. I. Title.
HQ663.9.S53 2012
306.0956 — dc23
2011036906

Book design by Brian Moore

Printed in the United States of America
DOC 10 9 8 7 6 5 4 3 2 1

To my wife Nada, daughter Laila, and son Malik
And to Jedeidet Marjayoun, as it was and will always be

CONTENTS

PART TWO: AT HOME

The true Vienna lover lives on borrowed memories. With a bitter-sweet pang of nostalgia he remembers things he never knew. The Vienna that is, is as nice a town as ever there was. But the Vienna that never was is the grandest city ever.

— Orson Welles, *Vienna* (1968)

Bayt Samara

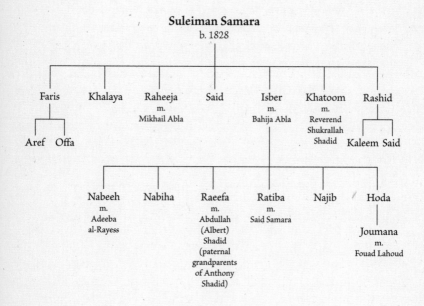

Suleiman Samara
b. 1828

Faris
Khalaya
Raheeja
m.
Mikhail Abla
Said
Isber
m.
Bahija Abla
Khatoom
m.
Reverend
Shukrallah
Shadid
Rashid

Aref   Offa

Kaleem Said

Nabeeh
m.
Adeeba
al-Rayess
Nabiha
Raeefa
m.
Abdullah
(Albert)
Shadid
(paternal
grandparents
of Anthony
Shadid)
Ratiba
m.
Said Samara
Najib
Hoda

Joumana
m.
Fouad Lahoud

Family trees contain only the names mentioned in the book.

Bayt Shadid

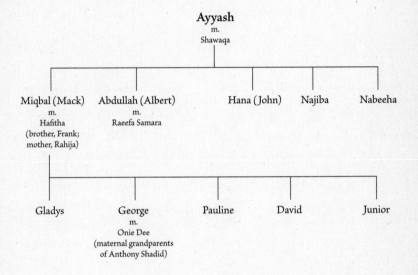

Ayyash
m.
Shawaqa

Miqbal (Mack)
m.
Hafitha
(brother, Frank;
mother, Rahija)

Abdullah (Albert)
m.
Raeefa Samara

Hana (John)

Najiba

Nabeeha

Gladys

George
m.
Onie Dee
(maternal grandparents
of Anthony Shadid)

Pauline

David

Junior

INTRODUCTION: BAYT

The Arabic language evolved slowly across the millennia, leaving little undefined, no nuance shaded. Bayt translates literally as house, but its connotations resonate beyond rooms and walls, summoning longings gathered about family and home. In the Middle East, bayt is sacred. Empires fall. Nations topple. Borders may shift or be realigned. Old loyalties may dissolve or, without warning, be altered. Home, whether it be structure or familiar ground, is, finally, the identity that does not fade.

I N OLD MARJAYOUN, in what is now Lebanon, Isber Samara left a house that never demanded we stay or enter at all. It would simply be waiting, if shelter was necessary. Isber Samara left it for us, his family, to join us with the past, to sustain us, to be the setting for stories. After years of trying to piece together Isber's tale, I like to imagine his life in the place where the fields of the Houran stretched farther than even the dreamer he was — a rich man born of a poor boy's labors — could grasp.

In an old photo handed down, Isber Samara's heavy-seeming shoulders suggest the approach of the old man he would never become, but his expression retains a hint of mischief some might call youthful. More striking than handsome, his face is weathered from sun and wind, but his eyes are a remarkable Yemeni blue, rare among the Semitic browns of his landscape. Though the father of six, he seems beyond proper grooming. His hair, apparently reddish, is tousled; his mustache resembles an overgrown scattering of brush. Out to prove himself since he was a boy, Isber would one day come to believe that he had.

By the time the photo of Isber and his family was taken, he was forty or so, but I am drawn more to the Isber that he became — a father, no longer so ambitious, parted from his children, whom he sent off to America to save their lives. I wonder if he pictured them and their descendants — sons and daughters, grandsons and granddaughters, on and on — moving through lives as unpredictable as his. Did he see us in years ahead, adrift, climbing the cracked steps and opening his doors?

At Isber's, the traveler is welcome, befitting the Bedouin tradition of hospitality that he inherited. The olive and plum trees stand waiting at this house of stone and tile, completed after World War I. The place remains in our old town where war has often stopped time and, like an image reflected in clear water, lingers as well in the minds of my family. We are a clan who never quite arrived home, a closely knit circle whose previous generations were displaced during the abandonment of our country decades ago. When we think of home, as origin and place, our thoughts turn to Isber's house.

Built on a hill, the place speaks of things Levantine and of a way of life to which Isber Samara aspired. It recalls a lost era of openness, before the Ottoman Empire fell, when all sorts drifted through homelands shared by all. The residence stands in Hayy al-Serail, a neighborhood once as fine as any in the region, an enclave of limestone, pointed arches, and red tile roofs. The tiles here were imported from Marseilles and, in the 1800s, suggested international connections and cosmopolitan fashionableness. They were as emblematic of the style of the Levant as the tarbush hats worn by the Ottoman gentlemen who lived in the Hayy, where the silver was always polished and the coffee came often in the afternoon. Old patriarchs — ancient and dusty as the settees — wiped rheumy eyes with monogrammed handkerchiefs. Sons replaced fathers, carrying on treasured family names. Isber was not one so favored.

In a place and time not known for self-invention, Isber created Isber. His extended family, not noteworthy, consisted of "less than twenty houses." His furniture, though expensive and imported from Syria, was as recently acquired as his fortune, and his house stood out not just because of its newness. It was a place built with the labor of a rough-hewn merchant whose eye was distracted from accounts only by his wife, Bahija. It serves as a reminder of a period of rare cultivation and unimaginable tragedy; it announces what a well-intentioned but imperfect man can make of life. Isber's creation speaks of what he loved and what sustained him; it reminds us that everyday places say much, quietly. The double doors of the entrance are tall and wide for men like Isber, not types to be shut in.

Isber, whose daughter Raeefa gave birth to my father, was my great-grandfather. I came of age with remembrances that conjured him back to life, tales that made him real and transported my family to his world, a stop gone missing on recent maps: Jedeidet Marjayoun. This is the way my family refers to our town, our hometown. Never Jedeida, never just Marjayoun. We use the full name, a bow of respect, since for us the place was the beginning. It was *bayt*, where we came to be.

Settled by my forebears, Marjayoun was once an entrepôt perched along routes of trade plied by Christians, Muslims, and Jews which stitched together the tapestry of an older Middle East. It was, in essence, a gateway — to Sidon, on the Mediterranean, and Damascus, beyond Mount Hermon; to Jerusalem, in historic Palestine; and to Baalbek, the site of an ancient Roman town. As such, this was a place as cosmopolitan as the countryside offered. Its learning and sophistication radiated across the region.

Yet lingering in small places is not in favor now; they no longer seem to fit the world. Yes, Marjayoun is fading, as it has been for decades. It can no longer promise the attraction of market Fridays, when all turned out in their finery — women in dresses from Damascus, gentlemen with gleaming pocket watches brought from America. At night, there are only flickering lights, which even a desperate traveler could overlook. In the Saha, or town square, there are dusty things — marked down for decades—for sale. No merchants shine counters, or offer sherbets made from snow, or sell exotic tobaccos. The cranky sheikh who filled prescriptions, if he cared to, is no more. The town no longer looks out to the world, and it is far from kept up. Everywhere it is scattered with bits and pieces, newspapers from other decades, odd things old people save. Of course, no roads run through Marjayoun anymore. A town whose reach once spanned historic Syria, grasping Arish in the faraway Sinai Peninsula of Egypt before extending, yet farther, to the confluence of the Blue and the White Niles, now stretches only a mile or so down its main thoroughfare.

Once, in this place, my family helped raise the cross and disturb the peace. We were known here, not for gentle natures or even temperaments, though we were among the town's first Christians. We walked these streets, played a role in determining where they would go. And then we used them to leave. Although our family tree still has olives on its branches, we follow the tradition of remaining *mastourin* (hidden, invisible, masked) when it comes to emotions, yet there are sometimes tears when we look back.

Isber's is one of the many houses left behind here, one of those we call *mahjour,* an Arabic word meaning abandoned, forsaken, lonely. The leftover houses—spindly, breaking down, haunted—speak of Marjayoun's lost heyday. For many who have walked by them through many years and wars and passings, they are friends. In their shattered windows, those who pass by see shiny panes and all that happened behind them. In the dark rooms they envision, not just scarred or peeling walls or dusty floors, but old acquaintances lighting lamps or stoking the coals of stoves.

The story of the town is written in these places; it is a history of departures. *I still think of them every day.* The houses of those who left are everywhere, walked away from. *There were letters for a while. She was my best friend.* Those who stayed remember those we lost. *We woke and saw that their place was empty.* In these broken-down rooms one can hear the voices of ghosts and the regrets of those who still recognize them.

• • •

Close your eyes and forget Marjayoun. The next thing you are crossing is the Litani Valley, over the mountains to Jezzine and then down the coast to Saida.

My aunts and uncles, grandparents and great-grandparents, were part of a century-long wave of migration that occurred as the Ottoman Empire crumbled then fell, around the time of World War I. In the hinterland of what was then part of Greater Syria, known locally as *bilad al-Sham,* the war marked years of violent anarchy that made bloodshed casual. Disease was rife. So was famine, created by the British and French, who enforced a blockade of all Arab ports in the Mediterranean. Hundreds of thousands starved to death in Lebanon, Syria, Palestine, and beyond. Isber's region was not spared. A reliable survey of 182 villages in the area showed that a fourth of the homes there had withered into wartime ruin, and more than a third of the people who had inhabited them had died.

This horrific decade and its aftermath provoked villagers, including my family, to abandon their homes for locations from South America to West Africa to Australia, as well as a few neighborhoods in Oklahoma City, Oklahoma, and Wichita, Kansas. What became an era of departures ended with more Lebanese living in the diaspora than within the boundaries of 1920, when Europeans parceled out the unbroken expanse of the Ottomans.

A green folder sits in my file cabinet. *Family Records,* it reads. Inside are citizenship and marriage certificates, my grandfather's discharge orders from the U.S. Army, my grandmother's story, written by one of her daughters, and a record of my grandfather's journey from Beirut to Boston aboard a ship called the *Latso.* Creased and folded in thirds are family trees from both sides of my clan, the Samaras and the Shadids. The first traces back to one Samara Samara, who was born in 1740 and emigrated in an epic exodus said to be led by women from the Houran of present-day Syria to the hills of Marjayoun. The other, much more complex, radiates into more than two hundred branches of names, insistently rendered in English and Arabic.

The folder also contains pictures. In one, my maternal great-grandfather, Miqbal, boyish-looking then, wears an ill-fitting formal jacket with an oversize white rose in his lapel. Other photos portray wistful ladies and men with handlebar mustaches and tufts of what appears to be quite unmanageable hair, all dressed as dandies in their Sunday best. There is one of the dry-goods store of an older Miqbal, where signs offered *High Quality, Low Prices.* But the English is uncertain: *Help Us, Weel Help You.* And the script is distinctly native, the graceful slope of Arabic, leaning to the left, imposed on the rigidity of Latin, standing straight.

The America that drew my family was a journey of seven thousand miles, and although mountain roads and voyages in steerage were treacherous, the hardest were those first miles away from home, away from faces that would no longer be familiar. By the time we arrived in New York, or Texas, or Oklahoma, or wherever, much was lost. "Your first discovery when you travel," wrote Elizabeth Hardwick, "is that you do not exist." In other words, it is not just the others who have been left behind; it is all of you that is known. Gone is the power or punishment of your family name, the hard-earned reputations of fore-bears, no longer familiar to anyone, not in this new place. Gone are those who understand how you became yourself. Gone are the reasons lurking in the past that might excuse your mistakes. Gone is everything beyond your name on the day of your arrival, and even that may ultimately be surrendered.

So much had to be jettisoned for the sake of survival. Emotions were not acknowledged when so many others had suffered more. There was only survival for these travelers and faces to recall until the pictures they carried frayed or no longer held together. Though none of us could summon its image, Isber Samara's house remained, saying his name and ours. It was a place to look back to, the anchor, all that was left there. To my family, separated or reunited, Isber's house makes a statement: *Remember the past. Remember Marjayoun. Remember who you are.*

PART ONE

RETURNING

WHAT SILENCE KNOWS

July 30, 2006

S LOWLY," THE TOWNSPEOPLE had cried to the man driving the bulldozer flattening what remained of their town. "Slowly, slowly." It seemed that I heard in their voices all the others of those I had known over the years who had lost their homes.

Some suffering cannot be covered in words. This had become my daily fare as a reporter in the Middle East documenting war, its survivors and fatalities, and the many who seem a little of both. In the Lebanese town of Qana, where Israeli bombs caught their victims in the midst of a morning's work, we saw the dead standing, sitting, looking around. The village, its voices and stories, plates and bowls, letters and words, its history, had been obliterated in a few extended moments that splintered a quiet morning. In the path of a bulldozer clearing the wreckage of lives was what would remain: a bag of onions, a can of beans, a blood-stained blue mattress, a teakettle, a photograph of a young boy, posing uncomfortably, backing awkwardly into manhood.

Slowly, slowly. The request repeated itself to me as, searching for some telling detail for another story to appear in the *Washington Post,* I noticed the fragrance of cedars and pines. Their smells seemed fresh and bracing, promises of renewal, until I discovered that the actual trees had been destroyed hours before.

I had arrived in Qana to see webs of wire dangling along the suggestion of a street. Some Lebanese believe that it was here, amid grape arbors, olive groves, and fig trees, that Jesus performed his miracle, turning water into wine. Yet on this summer day, olive trees with gnarled trunks perhaps a century old were split like toothpicks. A tattered Persian rug jutted out the back window of an old Chevy, hurled from somewhere by an explosion. As a donkey brayed, a terrified cat shot through the rubble while Israeli shelling thundered in the distance. Moments later, a rescuer rose from the ruins, back slightly stooped. Cradled in his arms was a one-year-old child, Abbas Hashem, the twenty-seventh victim of the bombing of Qana. A blue pacifier dangled from his green top. A bruise covered his forehead, and his tongue hung listlessly from his mouth. Behind him lay a book, *The Keys to Heaven,* the corners of its pages charred.

Most of the dead had choked on flying dirt and other debris. Their bodies, intact, preserved their final gestures: a raised arm called for help, an old man pulled on pants. Twelve-year-old Hussein Hashem lay curled in the fetal position, his mouth seeming to have vomited earth. Mohammed Chalhoub sat on the ground, his right hand broken. Khadija, his wife, and Hasna, his mother, were dead, as were his daughters, Hawra and Zahra, aged twelve and two. As were his sons: Ali, ten; Yahya, nine; and Assem, seven. "I wish God would have left me with just one child," said the bereft former father.

War had come home again to Lebanon, where, since World War I, it has been more familiar than peace. For eighteen days I had covered Israel's latest attack. With my fellow reporters I had followed a campaign deadlier and more destructive than any here since the Israeli invasion of 1982, which began an eighteen-year occupation. Israel had stormed in after Hezbollah — the militant arm of the Shiite Muslims in

Lebanon — infiltrated the heavily fortified Israeli border, killing three enemy soldiers in an ambush and spiriting two others away. In retaliation, the pretext for reprisals that are never proportionate, the Israelis unleashed a thirty-three-day barrage that destroyed entire villages and left more than 1,100 dead, most of them civilians. Their Merkava tanks plowed ahead as unmanned drones hovered, buzzing like swarms of insects. Most of the weapons were American — the F-16s and Apache helicopters, the Sparrow and Sidewinder missiles, the cluster bombs that left four million bomblets sown in the ground, waiting to kill and maim long after the war ended.

When my group said goodbye to Qana, I hoped for a quick trip on twisting roads through the hills and smoky greenery, flying fast away. Traveling in my old Jeep Cherokee, we headed to Tyre, where, upon arrival, we spilled out. What I noticed was not more death, though eighty-six faces lay in cheap coffins baking in the blazing sun of south Lebanon. What moved me was, in the corner of the scene, a row of women mourning in black. As the heat climbed, a few lifted their veils, moving cautiously, as if they feared a single gesture might discomfit the world. The women in Tyre did not flinch, did not speak; they did not ask that their sorrows be noticed. They were here for others, and as long as the caskets remained, waiting to be interred in the same gaping hole, the women would not depart. Their presence said that life was still sacred, that the loss of it mattered, even now. In the Middle East, the first lesson is the meaning of silence. In the silence of the women was faith.

In my silence there was my family, on my mind since this war had started. Maybe it is because my relatives are emigrants that I rush my departures, which I believe are best made early in the morning, in the dark, before babies cry out, or wives awaken, or callers from Bangalore demand payment of credit card invoices. I would rather say nothing and run. Better silence than words second-guessed across the globe. Grab the suitcase. Watch for the taxi's headlights. Smoke a forbidden cigarette. Go.

But the place I was usually headed to was no longer the one with which I had once been so enamored. The Middle East that had fascinated, preoccupied, and saddened me for decades was gone. I had first come to know it when I was in college and spending a summer in Jerusalem. Then and later, excursions took me from the crumbly Ottoman outpost of Suakin, on the Red Sea, to the oily sprawl of Riyadh and, stretching across the desert, to Sanaa. From that magical old town, with its toy-like houses of stained glass and white gypsum washing over cream adobe, I traveled on, to capitals of surreal modernity on the turquoise waters of the Persian Gulf. Especially compelling to me was something difficult to articulate but that underlay everything, an approach to life — an ease, an elegance, an absence of the unnecessary. Anything hurried, superficial, purely mercenary, or delusory was rejected. Central was a slowness allowing for the consideration of every choice. The state of the spirit, it is believed, reveals itself in small tasks, rituals — all the things that war interrupts.

Old traditions that represent values, daily habits that calm the mind, are not perpetuated when war stops time. Life goes unattended. What might have been lasting is lost. The old ways of the Levant have dwindled down here, as war — or the threat of it, or the wait for it, or the loss that follows it — has become a way of life. By the time I arrived in Lebanon in 2006, few cultures intersected here. Politics was refracted through unyielding religious discourse or more ancient affiliations, and identity flowed exclusively from them, irrespective of culture and language. It seemed we had been left with tribes bereft of citizenship. Home, united, as other generations had known it, had long been lost, though an older architecture still whispered of times glimpsed in broken masonry and solitary arches.

At 3:30 A.M. on August 10, 2006, a grim Thursday, the Israelis entered Marjayoun, whose name, in Arabic, means Field of Springs. War was about to hit home, and of course I wondered whether Isber Samara's house would be wrecked. I knew I had to go, and days later I set out to visit that stately old home perched at the foot of Mount Hermon, known to all here as Jabal al-Sheikh, the Mountain of the Old Man. The house had been abandoned for years.

The drive to Marjayoun should have been leisurely; it was just a few hours away on hilly roads to the market town of Nabatiyeh, then through the Litani Valley to Marjayoun. But en route from Tyre that day, no one could say what roads would collide with the war. We traveled an out-of-the-way, serpentine route, trying to locate a clear, safe passage, sheltered from the sky. They say you don't hear a missile before it strikes your car, but I listened anyway. I was trying to stay quiet. If I spoke, I would have almost certainly lost patience with someone.

Fresh from three years reporting the conflict in Iraq, I had been grazed by three wars and was suddenly covering — with the help of enough cigarettes to keep the Carolinas out of the meth business — the grisliest conflict I had tried to describe. The gray strands in my hair were not, at that point at least, signs of aging, but keepsakes from Ramallah and Baghdad, not to mention the final six months of my marriage. I think it was the last that left me with the most gray. The battles with my wife had been accelerating for what seemed like ages. My daughter's mother had been understandably obsessed with the lethal aspects of my career since March 2002, when I was shot by an Israeli sniper in Ramallah.

Even before I had heard his bullet that day, I fell, momentarily deafened and disoriented. A stun grenade, I thought at first. I couldn't move my arms or legs, but soon felt a sting on my spine, grazed by the round that was probably aimed at my head. "I think I was shot," I remember saying to my Palestinian colleague, seemingly hours after I had realized it, though only moments had passed. He lay next to me, desperately patting my body and looking for hemorrhaging.

The warm blood that soaked my dirty clothes felt almost soothing as I lay crumpled under a cemetery-gray sky. I recovered, for the most part. My wife did not. Our home was breaking, broken, finally broken up. By the time I arrived in Lebanon, I was a suitcase and a laptop drifting on a conveyor belt.

To get to Marjayoun, occupied by the Israelis, we sped north to the Litani River, but couldn't find a way across. The bridge was in ruins — like the makeshift road over the water, blown by bombing. Though not much of a river, the Litani is too deep to ford, even in summer. But someone spotted a temporary bridge, apparently erected the night before (probably for guerrillas to transport their weapons), and we managed to cross. We made it to the capital, Beirut, with no reports filtering in from our destination.

In Beirut at that time, one song, by Majida al-Rumi, was often heard: "O Beirut, Lady of the World." "Rise from the rubble like a flower of almond in April," Rumi sang, voice soaring. "Rise, O Beirut!" But driving toward the city, I knew the Beirut of tomorrow would never be the same as the one that, only recently, had been so buoyant. By the time we reached it, fuel depots were still burning at the airport, with columns of white smoke billowing over its seaside. After each guerrilla attack, Israel stepped up the bombings, systematically dismantling the capital's infrastructure, displacing hundreds of thousands and instilling a sense of foreboding, fear, and defiance in those quarters where militant politics merged with the deepest faith.

In homes that we drove past, residents watched on television the attacks a few miles away, which they could hear and feel. In the streets, the tire tracks of ambulances cleared paths through broken glass. The call to prayer echoed across suddenly deserted alleys. The occasional car that passed mine was usually headed for the Syrian border, the last way out of the country.

From Beirut, we crossed the mountains to Zahle and down the Bekaa Valley and its vineyards, always in search of gasoline for our cars. We plowed past abandoned checkpoints and over hilly, moonlit roads. There were no cars. No one was in the streets. The Lebanese soldiers had long fled, before the rotors of choppers beating the air and the drone of surveillance planes.

By midnight, we came close to Marjayoun but had to turn back. A crater was carved into one road at the spot where it bent around a hill of crumbling terraces and worn stones. Rocks blocked another road, impassable but by foot. The one path to the town was a sandy trail that arched over a deserted quarry, across a ridge, and into the valley near

Marjayoun. But the Israelis were still there, occupying my family's old town, and no one could pass for now.

From the roof of the house of a small-town mayor, who reluctantly offered us shelter, I could see the hills that drew my ancestors from Syria in their exodus many centuries before. Mount Hermon stood like a sentinel, no longer capped in snow. Marjayoun was a half hour away, but I knew the olive trees there were full, their bounty not yet ripe. Recalling the exploded branches of Qana, I wondered whether the fruit would survive until the fall harvest. Though shaped by history, Marjayoun had never faced the full wrath of Israel, nor would it, probably, but war's trump card is unpredictability: In Bint Jbeil, a day or so before, I had seen rubble swept against houses like snowdrifts. Flies landed on lifeless eyes, and hundreds had taken shelter in a hospital near a Crusader castle, huddled in the dark, having followed rumors of elusive refuge. Elderly women, their swollen, bloodied, and bruised feet wrapped in gauze, waited in basements lit with candles that shone on sweaty, shiny cheeks.

"O Lord!" cried sixty-year-old Saadeh Awada, leaping up from a tattered cushion. "God stop the bombs!" Her screams made the children cry harder, the heat seemingly to move closer.

"Shut up!" one man shouted from a crowded hallway, barely lit by sun.

Everything in war here becomes personal. Behind today's skirmish or bombing lies an event that happened to a family yesterday, or decades back. "You have to understand," said the Israeli novelist David Grossman, whose son Uri died in the conflict of 2006, "that when something like this happens to you, you feel exiled from every part of your life. Nothing is home again, not even your body." What befalls a trained soldier during combat between nations is one thing; what occurs at home — on our street, in our yard, and on our land, to family — is not the same. In Qana, those who died would not flee, would not leave their homes. That is what *bayt* means.

"Slowly, slowly," they had cried in that old sad town as the bulldozer crushed the remnants of lives. I remembered the chorus of broken dishes, shattered pieces. Would Marjayoun become another footnote of war? I was still days away, but the town, and the house that Isber Samara had built long ago, mattered to me. I wanted it to survive.

The first time I walked through Isber's door, only a few months before, I had not been impressed; in fact, I felt no connection, and the place was trashed. During the occupation, my family had abandoned the house. When I tried the door, the key didn't work. Finally, after a long struggle, it creaked to the right as a plume of dust came after me. I recalled intricate cobwebs worthy of some ancient archaeological relic.

After Isber died, the house had been divided in two. Upstairs was the family home, where my grandmother Raeefa was born and where her mother, Bahija Abla Samara (Isber's wife), lived until she passed away in 1965. Before Isber died, when the house was new, Bahija devoted almost every moment to its upkeep. Her embroidered pillows decorated the house; her beige curtains adorned the windows and arcade. Day in and day out, Bahija toiled on her hands and knees, polishing the marble floor. The house never shined, it reflected: Faces floated on the surfaces when guests visited, and flames from kerosene lamps glimmered in the glass.

On that first visit, I couldn't really appreciate Isber's house. I wasn't ready for an encounter with it or its creator. Yet tentatively I stepped out onto the balcony that had been my great-grandfather's favored spot. Not far off, in the foothills that surrounded me, interspersed among the crumbling villas and fallow fields, were the town's smaller springs: *al-Tini, al-Safsaf,* and *al-Shibli.* They had given Marjayoun its name: *Al-Ain al-Kabira* was down a sloping road to several nearby villages. *Al-Ain al-Saghira* was a short distance away.

My great-grandfather had been dead since 1928, before some of his children—Nabeeh, Nabiha, Raeefa, Ratiba, Najib, and Hoda—had reached maturity. I had never met him, but I remember trying to piece him together. This effort would continue for years—on trips, through letters, family journals, and stories. Whatever relationship emerged, it was Isber's land that was the catalyst. The beauty of Lebanon neither shouts nor declares. There is a gentleness to the landscape of hills rounded by age and terraces crumbling for centuries.

I scouted for Mount Hermon, where winter still reigned, but the peaks lay across the valley where the land gave way to the horizon. The picture went on and on; it seemed indefinite, like those Chinese paintings where the horizon recedes into the clouds so gradually and vaguely that it is impossible to clarify exactly where the land subsides. My eyes followed it, back and back, searching for the end and never finding it.

The trip back to Marjayoun had finally ended, but the questions that had driven me there remained, repeating themselves in my dreams: What was left? What survived?

Words can't quite re-create the smell of war. I have found myself trying to wash it out of my hair, off my fingers. More than once, I have run water over the soles of my shoes. On the previous afternoon, by the time we got there, the Israelis had gone, but tatters of smoke still snaked through the streets. I smelled war in the square and in every breath. I did not want to smell it at my great-grandfather's house.

The damage inflicted upon Marjayoun, as it happened, paled before the destruction visited on other Shiite towns, though Marjayoun had not had it easy. After a few fighters allied with Hezbollah — or claiming to be — fired a few rounds at an Israeli convoy, the Israelis struck back hard, firing randomly, cratering buildings downtown, in the Saha, which was gutted by fires. Samir Razzouk's shop in the town square was no more; gone was the lottery machine, the shelves stuffed with decade-old oddities hoarded by the merchant, a notorious pack rat. The trail of the Israelis' destruction followed the road they entered; house after house was left with craters and bullet holes. They occupied a few places, sometimes defecating on the floor.

"They came with tanks, of course," said the mayor of Marjayoun, Fouad Hamra, his glazed eyes seeming to lag behind his words. For those who weathered the attack — perhaps four hundred residents, mainly elderly — the Israelis were ghosts, mostly hidden by smoke, darkness, or ominous headgear. Until the soldiers left, the Marjayounis, hidden in their houses, spoke only in whispers. They knew the drill: Voices attracted bullets.

After interviewing the mayor, I set off to Isber's along a buckling asphalt road, ignoring work, calls, reporting, stories due. As I entered Hayy al-Serail, Isber's neighborhood, where clouds of flowers once barely broke the stillness, I wondered how I was going to tell the cousins what had happened. My relatives are people who can fall into an argument over the choice of a toothpick; or, at an unexpected moment, smile buoyantly and clutch your cheek; or imply with a glance that you will soon be languishing in prison. (Not unjustifiably.) They are naturally mercurial, passionate. In Marjayoun, the citizens who remember say it flat out: "The Shadids are crazy." I didn't think my family insane. I believed simply that their secret mission on earth was to drive everyone else to the brink and then say, of the decades of therapy necessitated by their complexities, "I am not paying for this!"

Sometimes my relatives tell stories, but God forbid you should ask for one. Information is not something they surrender easily, and the past is not something they are always willing to recall. Clustered around the same several blocks in Oklahoma City, my family are separated by only a few houses, and are never far from each other. No one has suffered the curse of drifting off alone. Together is the way these people, my people, have lived since coming to the United States. Together is the way they will die. Community is everything; home is everything. If you have lost your own.

Minutes after leaving the town square, I was at the house of my great-grandfather, but it was my grandmother Raeefa, who spent her first twelve years here, who was most on my mind. Hers were the rhymes brought to the bare plains of America from Lebanon and sung to children, including me. *Oh Laila, there are no eyes like her eyes and the magic in her eyes.* I heard these simple songs on that strange, complicated day as I sat on the step near where, less than a century before, Isber had lifted Raeefa up into the buggy that would take her away, past Mount Hermon and down the Litani Valley. It was before she became someone else, before Beirut, the ship, or the sea. It was before Ellis Island, or Mexico, or crossing the river to America. It was before Oklahoma with its cowboy belt buckles and women with red lipstick

WHAT SILENCE KNOWS • 13

under hair-dryer helmets. I wondered whether the ancient olive trees on either side of me were ones she might have glanced at as she parted.

And then, my eyes turning to what I had instinctively sensed was awaiting my inspection, I found myself wondering what my grandmother would have made of the half-exploded Israeli rocket that had crashed into the second story of her father's house, taking out a good chunk of wall before bursting into flame. The Lebanese stonemasons, hailing from Dhour al-Shweir, Khanshara, and Btighrin, who helped bring the home that Isber imagined into reality, considered the limestone used on the old house to be impenetrable, but with new technologies and old antagonisms in play, there is nothing war cannot crumble in a heartbeat.

A few hours after discovering the rocket, I returned and, with a borrowed shovel, started digging. The topsoil was poor and friable, ravaged by the elements. As I dug further, the rocks turned to pebbles, and deeper, the soil became rich and fertile. The olive tree I had managed to purchase cost me $4, probably too much. Its trunk was no thicker than a pen, and its branches arched no higher than my chest. But set in the hole, ten feet from the trees dating from my grandmother's day, this late arrival, I hoped, might somehow make the statement to my daughter and her generation that Isber's house, whatever its condition, remained a home worth care.

What I felt was *bayt*, and it led me to make a promise to myself, a commitment that I still cannot believe I honored after all these years. You see, I have not always been a man who kept his promises, and I have never been the type to stay home.

LITTLE OLIVE

August 10, 2007

I T WAS A HOT summer day, and it seemed I had traveled through the ages since the afternoon I discovered the rocket at Isber Samara's house. I bent down to touch the olive tree I had planted more than a year before. Determined, it had shown itself, yet my pride at its survival was tinged with disappointment: It had grown, but not as much as I had envisioned. All these months later, the little tree still looked inconsequential before the house of stone. The trunk seemed incapable of outlasting a stiff wind and was far from outgrowing its vulnerability.

When I was younger, I had pictured myself as the sort who plants trees that endure. I wanted to be a family man, a generous man, but there was always work, and like my great-grandfather Isber, I had to make my name. Others, I mistakenly believed, would understand the sanctity of my mission and have no trouble postponing their cocktail parties.

But there I was, still stunned by war, and shockingly, no longer young, or married, or with my daughter, Laila. I was, as fate would apparently have it, living alone. Not in Washington. Or Baghdad. Or Beirut. I was residing, more than a year after the war of 2006, in Marjayoun, where I planned to rebuild that stone house, Isber's home, our home.

There is a place in suburban Maryland where my wife and I had tried to come together, though apparently not as hard as we tried to fall apart. There is a room where a baby slept as I said goodbye, off on another errand of career-building, feeling guilty for leaving, but not sorry enough to stay. I had tried to keep a hand on home, but was a kind of guest star, and sometimes, as wars accelerated, I forgot the plot unfurling back there. On what had, at the outset, seemed a promising summer day, I had returned to our house to find that my wife and daughter had vacated. The lawn was mowed, the flowers were planted, the tomatoes starting to ripen, but inside, precisely half of everything was missing. It was a clean, surgical division, worthy of the woman I had married four years before, a doctor.

Everywhere these days, all around the world, it seems, there are leftover houses, rooms without warmth. This was not the world that Isber Samara had known when young, before war changed his life.

Isber and Bahija married in 1899, and the union was considered, with no sense of diminishment, sensible. Bahija's forebears were from Salt; her face was soft and without schemes. Money was not always on her mind. She had kindness and dignity. Of six sisters, she was known as the prettiest, but this was considered no distinction. All the sisters were passable goods, wives-to-be for regular men, full and proper as a row of birds.

Bahija made a discipline of truth and was a light to Isber, whose stories tended to serve whatever purpose he had in telling them. He was a man of intentions, not always expressed. She exuded something beyond force or strength — a serenity and equipoise. No syllable was extended frivolously. Without her, her husband might only have overwhelmed. Isber's eyes, bright blue, demanded; hers, green verging on hazel, re-

quested. Isber was rough and hard to rule. She was Marjayoun, proud
and educated, but quiet, basita. *She shuddered at outbursts and shunned*
gossip. Silence, she had been taught, was valued. It could keep anything
at bay.

I am not certain when I decided to move to Marjayoun. For a while
I waited for someone to save me from myself, yet — although friends
looked stricken when I shared my plans — no one arrived, so I charged
ahead, arranging a leave from the *Washington Post* and signing a lease
on an apartment that would house me until Isber's could take over the
task. My temporary landlord, Michel Fardisi, in his bulky red Mercedes, was a predictable image of corruption, with a satisfied belly and
thinning hair dyed to a black sheen. He resembled a bandleader in a
nightclub with unsavory associations, and his firing up of successive
Marlboro Reds came with wheezes too broad for parody. Here was
a man conscious of money in a town that, drawing on its mercantile
past, took pride in its *shatara.* The word's definition, at least the most
literal one, goes only so far; it suggests slyness, shrewdness, cunning, or
smartness. Popular use, though, has broadened the definition: *Shater*
connotes a hint of wiliness, a few shades of deception. To be *shater* is
to be undoubtedly cleverer than the next person. It is an enviable at-
tribute among those born for the squabbles of the marketplace.

Michel thought himself quite *shater.* During the rent negotiations,
his first offer, muttered to my friend Shibil, a native here, was that I
should pay the full $100 rent, even though the month was well in prog-
ress. After all, he said, he had kept the apartment open for me, fending
off tenants. (I had, before this meeting, known nothing of the place,
but let it pass.)

I suspected that he was simply gauging my gullibility. My Okla-
homa-accented Arabic, sprinkled with Egyptian colloquialisms, perhaps suggested I was dimwitted, as any foreigner here is considered.
So I protested, pointing to the apartment and its water-stained walls,
rickety door, mossy entrance, and weeds. It was hard to imagine that
it was in particular demand; it looked like a crib for a serial killer. In
the end, Michel seemed unwilling to invest his time in an argument.

He relented with a shrug, a poker player folding a hand never good but boldly played. He tried to suggest I had gotten a bargain, but I could not quite agree.

Later, early in the evening, watching for the man who was to deliver some furniture I had bought, I felt eyes peering at me from a window behind a fig tree. Soon an elderly woman came toward me. Looking me over, she seemed bent on learning whatever she could. Her face implied that she didn't much like my intrusion. No smile appeared, no sign of hospitality, none of the rich catalogue of greetings that Arabic boasts.

"How much rent are you paying?" she asked. "A hundred dollars?"

I nodded.

"Are you married?"

I shook my head.

"Are you bringing your own furniture?"

At that I paused, then nodded, squinting, a little unsure of her meaning. Even Shibil seemed baffled by the last question. Then the electricity went out, as it often does in Marjayoun, and she walked away, muttering to herself that the apartment was too small for me, and too expensive. A few minutes later, she returned. She had the same scowl, but in her hand was a candle so that I could light my new home.

Leaving the apartment one morning a few weeks later, I was planning to start the actual building. I glanced at Shibil's old shirt, stuffed in a gaping hole in the kitchen wall, but moments later was rumbling past ripening pomegranate trees and panoramas of Mount Hermon. All seemed well. I had decided to quit smoking and, breathing deeply, was pleased at not having lit up that morning. Yet there was something slightly disturbing beneath the surface. Bishara, a friend of Joumana, one of my last remaining cousins in Lebanon, had promised to bring workers to begin the reconstruction — no, that would be too modest: the *reimagining* of Isber's house. Not just workers, either. Artisans! Masters! I had been prepared for Olympians of Construction. Bishara, tinged with the ambition of an Albert Speer, had enthused operatically over the plans that they would soon execute. His descriptions were overtures featuring the sounds of walls falling, and unfolding arias devoted to the creation of new bedrooms. (Five, he convinced me, were necessary, as he said coyly, "A man never knows.")

Stepping into the spotlight for what would be a lengthy solo, Bishara would harmonize with himself as he led me through imagined hallways, passing new bathrooms with shiny, astonishingly expensive fixtures, a dining room fit for state dinners, a gargantuan kitchen for never-ending meals, and a calming sitting room where I could recover from my nervous breakdown — an event certain to follow my perusal of the bills no one mentioned.

Upstairs, according to the Master Builder, we would essentially create two new floors where there had been one. A new, somewhat elegant balcony perched over the *liwan* (entry hall) would majestically open to the triple arcade of windows and the view of snow-capped peaks beyond. In Bishara's vision, the roof's red tiles would remain, as would the marble in the *liwan*. So would the century-old iron railings, or *darabzin,* on the balconies and over the windows. Perhaps some of the beautiful tiles — that echo of a Levantine past — could be salvaged, and the stone could be cleaned. But most of everything was to be replaced. Reviewing the plan, I shuddered. The intense precision of the lettering, fit for a ransom note, convinced me that I was dealing with a psychotic.

And did we *really* have time for all this?

I suspected that my year's furlough from the paper would dwindle to a close before we had finished the installation of a working commode.

That morning, I walked across Bahija's overgrown garden, strewn with clumps of dead weeds and the detritus of forgotten gates and walls, knowing that we were starting, or at least hoping we were. Then I headed to the house, expecting to find Bishara's crew. But the place was almost deserted. Only Fouad Lahoud, our engineer and the husband of Joumana, was laboring, rather unindustriously — quite unindustriously, actually. There he was, chipping away at decades-old plaster with what appeared to be a metal bookend. As unflappable as I was temperamental, he smiled. It was meant to reassure me. It did not.

Preparing myself, I realized I was not quitting smoking that day.

Nor probably the next. Bishara, Fouad explained, the man of lofty visions, was still working on a project in the Bekaa Valley, a few hours away. As for his crew, they had returned to Syria, fearful of the deepening political crisis that pitted the government against Hezbollah, its opposition. It was paralyzing the state. In the capital, there was deadlock over the choice of the next president. Rival television stations derided each other's supporters as militiamen and hired guns. No one seemed to think resolution would come without some sort of violence, somewhere.

We'll never start, I thought. What would Isber have done? I lit a cigarette.

I had left for Marjayoun about a month before that day when construction did not begin, driving from Beirut in my green Jeep Cherokee. My provisions included a loaf of bread that Lebanese call *toast,* five cans of tuna, cheese, bags of roasted peanuts known as *kri-kri,* almonds, two bottles of water, asparagus, hearts of palm, pickled okra, and coffee, along with a mug. A bottle of Glenfiddich gurgled in my bag, along with clothes and sandals.

There was no address; Lebanon lacks the precise Zip Code sense of the United States. Directions tend to be anecdotal, even in peacetime: Pass a crumbling lighthouse, turn right at a revolutionary's portrait, left at a pasha-like statue, near a rock-studded wadi no one knows the name of. Roads tell a story, and though some details were new, the landscape was the same as my family had seen years back as, leaving Marjayoun, they headed to Beirut and America.

The Shadids were among the first to leave Marjayoun, joining others who fled, starting around 1894. After the murder of a sheikh from Metulla almost led to an attack on Marjayoun by Druze tribesmen, Ibrahim Shadid, a short, wiry man with a remarkable handlebar mustache, decided he had had enough of vendettas in a land too often filled with them. He left, making the monthlong journey to the United States by donkey, boat, train, and finally on foot — from Beirut, via France, to New York, and then to a town along the Red River and the border of

Oklahoma. He ended up in Sherman, Texas, a prairie town named for a hero of the Texas revolution and best known for hosting Jesse James for part of his honeymoon.

His brother Ayyash stayed behind. Yet when Ibrahim's other brother, Shehadeh, decided to join Ibrahim in Texas, Ayyash sent along his oldest son, Miqbal. Fourteen years old, with a round face on narrow shoulders and a cross tattooed on his wrist to ward off the evil eye, Miqbal heeded his father's wishes. He would never see his father again, although his relatives, trickling into the new country, would bring occasional word of his family, including the trials of Ottoman conscription and World War I. Together, Miqbal and his uncle Shehadeh began working as peddlers, migrating to Oklahoma, a territory that had just won statehood. Flush with oil, it was exuberant and lawless, as much a frontier as Texas.

Along the wadis I passed the day I drove to Marjayoun, empires had collided — Hittites, Arameans, Assyrians, Babylonians, Romans, Byzantines, Arabs, Seljuks, and Crusaders. With those empires came kings: Nebuchadnezzar, Tiglat-Pileser, Benhadad. More followed, from Amr ibn al-As to Saladin, the Kurdish nemesis of those Crusaders. Their heirs were the Ottoman beys and pashas, British field marshals, and French high commissioners.

The Ottoman state had begun as a negligible emirate led by a chieftain named Osman in the fourteenth century, but the Ottomans' sway over what would become Lebanon commenced later, under Selim I, whose feared janissaries conquered Egypt and Syria. At its apogee, under his successor, the immodestly named Suleiman the Magnificent, the Ottomans' empire was the most invincible, the wealthiest, and perhaps the most advanced domain on earth. It called itself the Eternal State. Spanning three continents and more than six centuries, it was Islam's equivalent of Rome, reigning over much of the Middle East, North Africa, and the Balkans.

These days, no one would long for the Ottoman imperium. Cited often — if the era is recalled at all — are the massacres perpetrated and the discrimination faced by Jews and Christians in taxation and commerce. Equality was unknown. Yet across those centuries, the Ottoman Empire bound together a remarkable tapestry of ethnicities, religions, nationalities, and languages that, unencumbered by borders, comprised a culture far greater than its individual parts. It survived as it did because of its pluralism and its own notion of tolerance. Albanians and Greeks, Armenians and Serbs, Arabs and Hungarians served the government; Christians filled the ranks of its janissaries; Muslims and Christians, Jews and Samaritans, Circassians and Arabs, Armenians and Kurds inhabited lands not too distant from Marjayoun.

When Muslims and Sephardic Jews were expelled from Spain in 1492, the emperor of the Turks sent his fleet across the Mediterranean to save them and settle them in imperial domains where Islam was never too orthodox, influenced as it was by tradition, mysticism, and even the Christians with whom it lived. No one was ever really Ottoman, a fact that some connect to the eventual failing of the empire. But since its fall after World War I, never has such a mosaic of cultures existed over such a breadth of land.

The war would be the end and the beginning. Even now, the most elderly recall stories of those last breaths of the empire — the *seferberlik*. It was the Ottoman name for the draft, but it came to mean something more: all the famine, terror, and disease that took lives and drained spirits in those years. Mere normalcy seemed disinclined to return. In Marjayoun, the only stories going around in these harsh days were grim tales of neighbors clawing through manure and chicken droppings in search of a morsel of food. At night, mothers and fathers sneaked into fields, cutting the tops of wheat, grinding the raw grain into an almost inedible sustenance for their children to eat. Others tried to eat grass. In the old Ottoman strongholds, people raided warehouses, looting them of loose kernels, olive oil, ghee, sugar, and onions. Cries of *juaan* — I am hungry — rang through the squares of Beirut at night. Epidemics of typhus and malaria preyed on the weak. Locusts attacked a countryside beset by plague. The dead were found in gutters, in clothes that barely concealed paper-like skin. Stories of insanity were routine.

The French declared Lebanon itself in 1920, fitting within its newly decreed frontiers eighteen religious sects—among them Shiite and Sunni Muslims, Greek Orthodox and Maronite Catholics, Druze, Armenians, and a handful of Jews. None agreed, or ever would, on what the country should become. So began the crises, which deepened after 1948, when Israel was created. The civil war raged from 1975 to 1990, and other conflicts have since inflamed rivalries and bloodied lives.

The old road to Marjayoun now passes through land demarcated with the colors, banners, and portraits of those conflicts, all of them abbreviations of reason. Yellow stands for Hezbollah and the territory of Shiite Muslims, the largest of Lebanon's sects. Portraits of an assassinated leader, killed in a spasm of flame and force on a bend of Beirut's corniche, denote the turf of Sunni Muslims, second only to the Shia in number. Flashes of green, red, and orange announce the irreconcilable Christians, their schisms myriad. Some retain deluded notions of supremacy, church and nation falling impossibly together. Each of their leaders has his title—the *sayyid,* the doctor, the general, the professor, the sheikh, the teacher, or the *bek.* In its labyrinth of contests, every community faces a future it deems existential. None has a guarantee of survival.

When I had arrived in Marjayoun that first night, I had accepted my friend Shibil's scotch and spare bed. Shibil is loquacious even if miserable; talk is his juggling act and he uses it to divert attention away from what he doesn't want to show. From his windows, the Israeli frontier is visible. Israeli checkpoints run along the ridge of Mount Hermon, part of it occupied by Israel since 1967. Beyond is Syria and ridges descending into Jordan.

Metulla, with its bold and steady illumination, taunted us as Marjayoun's lights flickered on and off while we drank. Marjayoun's power is rarely continuous. Accustomed to Baghdad, where lights reveal mountains of garbage and rivers of martial concrete (but only for twelve hours a day), I am taken aback by persistent utilities. The current that night was almost flirtatious, promising electricity then threatening to disappear in blackouts that lasted for hours.

Before I could point this out, Shibil interrupted: "The prostitute has gone," he proclaimed rather drunkenly as the lights failed once more. Darkness reigned, longer this time. When we were able to see again, Shibil exclaimed, "The prostitute has come."

Shibil cannot be extricated from Marjayoun, where his family, who once counted, is no longer recalled. In uncertain health, he tosses sleeplessly most nights, enraging and then reenraging himself as he sorts through his pile of grudges, imagined slights, and never-ending quarrels. They are what he has to prove his life is going on. He lives in a ghost town that survives mainly in memory, but to him it is the world.

Our first meeting, more than a year before, had been memorable. After some hesitation, he had invited me to his house, shrouded in a grove of olive trees and fruit-bearing cactus a little ways from the town's main road. It was a centuries-old stone family mansion turned chaotic college dormitory. Clothes were draped over every piece of worn furniture; his cabinet was crowded with vitamins and half-empty liquor bottles. Dusty cassette tapes were stacked next to a stereo with a turntable, and an astrology book lay open on the coffee table, with pages dog-eared. "We are good communicators, Geminis," he told me. It made his marriage to a Sunni Muslim woman difficult, though. "She's Cancer, I'm Gemini," he said by way of explaining their divorce.

Outside sat his white 1971 Mercedes. He always parked it elaborately, pulling in and out, reversing and moving forward, until it was just right.

"I always park ready to go," he said, and it was true. His car always faced downhill.

I asked him why.

"Escape!" he said, and he laughed hard.

A few days later, he would offer me a variation of his plan as advice.

"Coffee without cardamom is like a bride without her gown." This is a loose translation of the words that my cousin Joumana spoke as we shared coffee at a neighbor's house. Other visits had made me somewhat familiar to the few faces scattered around town, and during these early days in Marjayoun, invitations for coffee were issued furiously. As was the custom even in Bahija's day, the bitter drink is served in cups shaped like outsize thimbles. Poured from a blue, long-handled kettle, this batch was delivered on a silver tray with gold trim.

As far back as 1716, a visitor was said to have found remarkable the amount of this substance consumed by Marjayoun's people. He attributed it to their Bedouin origins — whether Christian or Muslim, they had retained the generosity and hospitality of a nomadic tribe. "I did not find in Jedeida one single hotel or café, but every house in Jedeida is a hotel," the visitor recalled. (He was struck, too, that after thirty days as a guest in one of those homes, in perhaps another Bedouin legacy, he never once saw the face of the family's women.) Nearly three centuries later, coffee in Marjayoun remains ubiquitous. "The most important thing in Jedeida is coffee," one friend claimed. "If you have a guest without offering coffee, then he is not a guest."

As Bahija knew so well.

In the land of Bahija's era, there were ways that women had learned to bring peace to the day. At this time, in climates of extremity, there were rituals and contemplations, tea and proverbs; there was the soothing quality of the garden rows, the quiet satisfaction of fabric properly folded, the stirring of sauces long to thicken. Every moment offered something to draw the eye, mind, and heart away from sorrows.

Each morning after her marriage, Bahija Samara, a dedicated creature of habit who considered herself the finest coffee maker in Marjayoun, woke at 4 A.M., never later than 5, to make Turkish coffee. On the patio she lit charcoal in an oven called a kanoun. *Once it was hot, she set down the long-handled kettle, known as a* rakwa, *and heated the water. Once it boiled, she drew it back, pouring in spoon after spoon of coffee grounds, as if it were a ritual. Then she put the kettle back on the oven and stirred the grounds in the boiling water with a teaspoon. The practice, taking more than an hour, was meditative, and the coffee reached a boil while she stirred. Its foam rose then receded as she pulled the* rakwa *back, only to rise again, until the taste was perfect, evidenced by the* wijh, *the thin layer of foam whose consistency and color she knew intuitively.*

Since I was single, friends warned me not to partake of coffee at the home of a single woman more than once; by the second time, the priest would be waiting to marry us. There were dark, somewhat bizarre suggestions of women performing black magic with coffee to lure an eligible man. (One friend blamed the unexplained for his brief marriage.) All had mastered the ritual of serving with hundreds of years of accumulated practice bringing grace to every gesture. According to custom, coffee or tea would appear only after sweets, fruit, or perhaps a mingling of walnuts and raisins. Chocolate usually followed. Guests were served first, family and hosts last. As we sat that day when construction did not begin — Joumana, her husband Fouad, Bishara, and me — I enjoyed a satisfied moment. Bishara was served first, coffee with its cardamom. Then came Fouad, followed by me.

Our conversation naturally moved toward the neighborhood and its rather unsettling quiet. Our host, Wadia Dabbaghi, was the wife of the headmaster of the Marjayoun National College. She lived across the street from the Samara house and named for me those who remained on the block: she, her husband Maurice, and a tailor who lived a few doors down and usually summered here. The rest, she had told me earlier, were either studying in Beirut or working abroad — in America or the oil-driven sprawl of places such as Dubai, where women outfit shrieking infants in Versace and all the mirrored surfaces are strictly self-reflective.

"You are the fourth," Wadia said, the faint hint of her smile suggesting that such a small number was, for her, not so happy a thing. She had no desire for solitude.

"I have a bad feeling about it," said Shibil, who had wandered into the gathering as the topic of rebuilding my grandmother's house was broached. "I hate to see you screwed," he continued. "This place doesn't belong to you." Why not buy land and build a house? Or buy a house that is yours alone and fix it up? "This house belongs to so many people. It's not yours."

He muttered an old saying in Arabic: "A sliver of land can wipe out its people." Then he nodded. In the days that followed, I wondered whether Shibil and the other naysayers weren't perhaps right. Their worry, simply put, was: I did not own the house. It was not legally mine. Disputes among heirs often blocked restorations here. Protracted family squabbles left many relics abandoned, like forgotten imperial outposts. I wasn't certain how I would fare in battle.

I was raised with an innocence at odds with the experience of my pragmatic Arab ancestors. To be born in these parts is not only to know loss and rumination, but also to savor the endless pleasures of discord. It is to feel, and often feign, useful rage. Anger diverts attention; as a ruse it can blur the facts of a losing argument or disguise one's true motives. Theater, at the negotiating table or during a midmorning's market dustup, is part of the action. Family battles here are freighted. Tales of the choleric House of Shadid do not necessarily inspire aunts and uncles struggling to reform tempers in the air-conditioned Oklahoma suburbs. When I discovered Hayy al-Shadadni, a neighborhood down the road from Shibil's that once bore our name, I anticipated angry gangs of diminutive men engaged in nighttime vices and excessive fist-shaking.

Anis Shadid, a mayor of our hometown long ago, knew how to cultivate the appearance of propriety. His tarbush, the once popular Ottoman-style hat, was always precisely placed. His suit was groomed, his leather shoes always bore a shine, and his cane was accented with a glimmering sterling silver handle. Yet nothing concealed the steam rising from his temples, the absolute refusal of any plea to reconsider or impose moderation, particularly when it came to his lusty appreciation of the water pipe.

One Shadid, frustrated by his schooling, buried his books in the garden to grow a tree of knowledge. In a show of loyalty or affection, another Shadid tried to get a passport for his pet under the name Bobby Shadid (*bobby* being another way to say dog in Arabic). My altogether more sane grandfather, Abdullah Shadid, soon-to-be-husband of my grandmother Raeefa, arrived in America and roamed the Texas oil fields before heading off to Detroit's factories and auto plants, where he and other Lebanese — Syrians or Turks, they were called — fought for workers in union battles. Oklahoma, more bucolic than Detroit, was his next destination. Here was respectability, never especially hoped for, arriving in the form of a dry-goods store with dusty shelves. Abdullah bought the business from a man who opened another establishment down the road. Furious at the deception, he and a relative sneaked out at night and, playing by the rules of village vendettas, burned it down.

What family in what nation fails to complicate divisions of money, houses, or land? It isn't just greed that breeds contention; there are also the grievances, hostilities, and rivalries of childhood. Everything, of course, is intensified by the frenetic but pitch-perfect negotiating talents on view in these lands, even among family. Imagine an endless, increasingly infuriated procession of swiftly modifying positions. Imagine the rampant muttering of threats with increasing ardor.

And so we begin. Any property in Lebanon — house, land, or those collapsing stretches of splendor that suggest the confluence of both — comprises 2,400 shares. Do not ask where the number came from. Someone will tell you, with utter confidence but no regard for accuracy. And then will come the correction of that figure, and then the voices will grow louder, more aggrieved. Within a half hour's time, a friendship or a marriage will be threatened with extinction. Documents will be torn or thrown. And then . . . the slam of the door? Not quite. The disputant, having stalked away, will reappear with further affronts. Time, eddying here as it does, is immaterial. Those with particularly effective memories like to tell the story of a man whose father was killed, inciting a vendetta. After forty years passed, his neighbors asked why he had not yet exacted revenge. "It's still early," he would tell them, ever so calm, patient.

This was the situation: The family's house was still, fittingly and according to the deed, in the name of "the heirs of Isber Samara." These came to twenty-three far-flung, combustible cousins, spread across Lebanon, Brazil, and five states in America — Oklahoma, Kansas, Missouri, Iowa, and Florida. Each of my father's generation had inherited 104 shares, give or take a few tenths. I could boast no more than 35 shares, and only if my portion of 34.78 was rounded up.

My ambition to rebuild the house was considered foolish and rash by my new neighbors, not to mention reckless, dangerous, and alto-gether "American."

Despite the anxieties and loud warnings, I remained determined. Joumana and I had agreed to do this for the sake of the family, or some notion of it, a hint of the martyr in our vow. We would leave owner-ship as it was. I suppose I should have talked to the stateside heirs and sought their blessing and, more correctly, their consent. I admit, though, I never did. My cousins' distance and long apathy had given me carte blanche.

I made my way back to Isber's house as the sun eased itself down to-ward earth and a soft, hesitant light filled the cracks and crevices of the homes I walked past.

Isber's, nearly a century old, was a formidable two stories of re-doubtable materials. Once utilitarian and still inescapably elegant, its stones ascended row after symmetrical row, thirty-five in all, ending with the *armid,* an Arabic word borrowed from Turkish that refers to those red tile shingles of a Levantine and cosmopolitan past. The roof was shored up by wood arches shaped like a sturdy violin's scroll. The two balconies were girded by finely wrought but rusted iron railings, perched over a burly plum tree. The larger of the balconies framed the most graceful feature of the house, the triple arches, buttressed by two marble columns. There is perhaps no feature that more force-fully evokes the notion of a traditional Lebanese house than the triple arcade, though inspiration for the design dates back nearly three mil-lennia, to Roman antiquity.

In Bahija's later years, the downstairs had been rented, sometimes to two families, and in the years that followed, the house did not escape scenes of darkness. When the civil war erupted in Lebanon in 1975, looters plundered the house, tossing delicate furniture into the yard and making way for squatters. During the long years of Israel's occupa-tion, a man with a strange, insidious air moved in upstairs. Some said Albert Haddad was a run-of-the-mill informer for the Israelis; others believed him a paid agent of the Mossad. He terrified everyone, so much so that stories of this sinister figure made their way to America. The carcass of a dog, still chained up, was found putrefying on the porch when Haddad left with the Israelis in 2000. Intelligence operatives — the sadistic, brutal kind used in Iraq and Syria, where they were known as the Mukhabarat — often work with attack dogs, savage animals used for, among other things, interrogation and torture.

I walked through the garden past scraggly pink geraniums, struggling without water, planted along the stone wall by a neighbor chagrined at the house's dilapidation. The door, shackled with a gold-colored lock that read BOXER, was a flimsy contraption of iron, wood, screen, and wire, hinged like a barnyard gate. Like all the entrances, it was battered and missing something: a lock, handle, or pane. At first the key I had been given didn't work, though I had used it on my previous visits. Then, as the door creaked, a plume of dust rose, and I saw shafts of light falling through the murky, disheveled room. Stretching across the threshold were cobwebs, glowing as they caught the glint of an afternoon sun.

Most of the downstairs looked looted and reminded me of scenes from Baghdad days after Saddam's fall. A faded picture of a stout model from an Arabic-language lifestyle magazine, her hairdo a decade old, was stapled to a beam. Wires dangled from the ceiling, as did light fixtures, though most bulbs were missing. The shoe prints of a dozen different feet left their mark in what suggested a hasty exodus. Yet gradually I began to see the underlying structure of the place, the almost hidden touches. There were no sharp angles or dead ends. As I walked toward the smooth stone stairs, I noticed some ornate Italian tile peeking from beneath all the dust. I was immediately drawn for reasons I can't explain. I am no aesthete, but I knew that the tiles were called cemento (though they were known these days as *sajjadeh*, Arabic for carpet, a name suggested by their repeating colorful patterns). Through the dirt I could see only black and white, but I suspected other hues lay hidden beneath all those years.

During my many visits to the Middle East, I had paid little attention to the cemento I had walked on countless times. It was there, I am nearly certain, on the floor of the Feisal Hostel — across from Bab al-Amoud, the Damascus Gate, in Jerusalem — where I stayed for some months as an angry eighteen-year-old student, working at an English-language Palestinian weekly during the first Intifada. I seemed to remember the tile later, when I was reporting in Cairo, in Midan Talaat Harb, on the floors of Groppi, Café Riche, or the Greek Club, landmarks of a downtown of Parisian ambition and Egyptian nostalgia. I remembered designs and colors on the floors of houses, shops, and restaurants I visited in faded quarters of Beirut and Nablus, Damascus and Aleppo, always wanting to lay claim to something that was never mine to take.

But it was the cemento here in Isber's house that drew me. At the foot of the stairs to the entrance, next to a slab of stone, there were twenty tiles, probably no more than an afterthought, as the space where they were laid was quite small — not much more than a few feet across, even fewer feet wide. The tiles had neither symmetry nor design (especially compared to the much larger, more carefully ordered displays I would discover on the floor above). Always frugal, my great-grandfather was never one to waste tile, and he had probably used leftovers here as a kind of accent. I could not stop staring. I was taken with their intricacy and the richness of the hues they promised. As I swept the dirt with my boot, I glimpsed colors that had resisted a century of wear — soft yellows, purples, and greens I had not previously encountered.

A week after learning that Bishara was still busy in the Bekaa Valley, Fouad dropped by my apartment on a Saturday morning. With a gentleman's sense of reassurance and an unexcitable, distinctly Lebanese confidence, he convinced me that all was not lost. He had, he proudly announced, been cogitating. There was a way forward after all — namely, with his brother Armando. Together, they would straighten out the design problems and rescue me. It seemed that Armando was also an architect and that Fouad, determined not to raise false hopes, had already taken Armando by the house so he could survey what needed to be done. Armando, thank the Lord and everyone else, had not been put off.

Here was their idea: They would take over the project, as engineer and architect. We would scale back Bishara's too elaborate, too time-consuming, too expensive plan and hire local labor. We would do what we could do, given the limits of time (a year) and money (never enough). As this was the first time I had heard anyone besides myself show the slightest concern for the financial outlay, I was ready to sign.

"It is going to be full of surprises," Fouad said as I lit another cigarette.

Fouad and Armando would come once a week, but we would also need a foreman to oversee the day-to-day work, unsnarling complications that inevitably tangled up construction jobs. Maybe our neighbor, Fouad said, knew of a potential candidate. I thought for a moment, then remembered Hikmat, a proud, strapping man who lived in a manor-like house down the road. Hikmat Farha was among the first men I had met in the town, perhaps because he was hard to miss — broad-shouldered with the obligatory mustache, a handsome hawkish nose, and thinning hair combed back. "You must be the tallest man in Marjayoun," I had told him then.

"The second-tallest," he corrected me. Apart he stood, and I trusted him.

Luckily, Hikmat did have a prospect for the job, a friend of his family's who, before the civil war began in 1975, had built the Farha family's gas station, a behemoth of concrete, at the edge of town. Hikmat called the man's son and asked him to send his father, who was called Abu Jean, but an hour brought no one. When I called again, Hikmat said that Abu Jean had come, waited an hour, given up, and returned home for coffee. The idea that he couldn't find us, right outside the house, in view of the street, was disconcerting. I set out to fetch him anyway, as nothing is too far away in Marjayoun. As it happened, Abu Jean recognized me before I did him, since I was that rarity in this town, a stranger. He waved from his balcony like a ship's captain sighting some long-lost compatriot and inviting him aboard his vessel.

Could he come over now, I asked, though this drew a blank look. When I repeated the request, he peered back, still seemingly not comprehending. Louder, I repeated the words. Finally he nodded, pleaded for a moment, sipped his thick Turkish coffee as he finished a previous conversation with a guest who peered suspiciously at me, then came down to the street. With few words, we were soon on our way, at first shuffling at the stately pace he determined, then, thankfully, catching a ride the rest of the way to the house.

Abu Jean, at seventy-six, defied nature with his full head of hair, combed back and still mainly black. His face was leathery, swarthy, and furrowed by age, the product of years under the sun, but it was as taut as a younger man's. His chin was sculpted.

I would come to learn that the half-finished cigarette — Cedar, a cheap Lebanese brand — that hung from his mouth was one of his trademarks. Decades of nicotine had stained his mustache. Although only of average height, he seemed extraordinarily strong but also, despite his intense masculinity, a kind of diva. He was, after all, a *maalim*, a hard-to-translate designation that, in the construction trade, can mean professional, expert, or master — as in master carpenter, master stonemason, master electrician.

Abu Jean was a builder, but more precisely, a *maalim* of concrete. He brought more than a half century of experience to his profession, a point he would make to me repeatedly — hourly, it seemed some days. As I would come to learn, he made his own rules, casually disregarding anyone, everyone, particularly other engineers, and most particularly younger ones. An artist, he called himself. The equivalent of sixty artists, he would add, voice rising. Abu Jean was not shy. That was simply the way it was. Once inside the house, he raised grandeur to a new height, speaking of the original builders as if he had been privy to their every thought and gesture. "This is how they built the world," he declared, patting the walls of the vaulted room. "They built walls like this," he said, pointing, apparently by accident, not to a surface impressive for its solidity, but to a hole where scraps of paper fluttered in the breeze.

At one point, I mentioned that we might need someone who could clean the stone and repoint the mortar. "Do you know people who can do this?" I asked.

Abu Jean, who tended to remain almost theatrically engrossed in his own consciousness, had a habit of not answering questions, for whatever reasons. Even if he deigned to consider the inquiry, he avoided a hasty response, instead offering vague reassurances.

"You should be relaxed," he said, clasping my shoulder with an iron grip. "Don't worry." When I told him I had to finish by May, Abu Jean counted the months, slowly, finger by finger. It was now August. We could work in September, October, and November. Not much would be done in December, January, and February, he said, but we could start again in March.

He looked at me and declared, with utter certainty, "We'll have it wrapped up by then."

"God willing," he added, repeating the phrase three times, as if to suggest a special covenant with the divine.

The next morning, Fouad was waiting at the house with Abu Jean. The crucial business that remained was not a small item, and I had lain awake the night before, watching a cockroach parade and agonizing over that haughty disrupter of so many romances: filthy lucre. In other words, Abu Jean and I had not agreed on his fee, and that morning I had a sense of foreboding as we exchanged pleasantries. Soon enough, the conversation turned to money, and Abu Jean once more returned to the theme of his six decades of experience. Then came oaths attesting to his honesty. "Just as there's a God in the sky" began one.

His soliloquy continued, flowing like a river of superlatives or defamations as we roamed from room to room, upstairs and down, and then we went out to the olive trees in front, where Fouad appealed for me to intervene.

"Abu Jean, I'm not Lebanese," I said. "I'm American, and I do things a little different. I have a specific budget, and I need to know how much I'm going to spend."

The pleas were to no avail.

"Let's see what the work costs, whether it's a day, two days, or three days. Let's wait until we put the pen to the paper and record what we spent," Abu Jean told me. The theme throughout the half hour of conversation: "In the end, you'll pay me what I deserve.

"Don't worry," he said, smiling, "and tell Hikmat that he doesn't have to worry."

These assurances were sprinkled with pinches of my shoulders, slaps on my arms, and smiles. He gazed upon me as if staring at a newborn grandson. To keep the mood light, Abu Jean threw out a few honorifics, at one point calling me *effendi,* a Turkish word for sir, and Fouad *sayyid al-ikram,* a term probably not used seriously in Beirut in a generation.

When Fouad finally got angry, Abu Jean named his price. "Seven hundred dollars," he announced. "Around seven hundred dollars." It was the equivalent of about 1,050,000 Lebanese lira.

Fouad turned to me. "Are you okay with that?"

I shrugged my shoulders as Abu Jean declared, to no one and all the world, "Everyone is running after money, and they end up losing their humanity."

I have to say, I liked his mind.

· 3 ·

THREE BIRDS

HISTORY IS SOMETIMES *written to buttress the myths that underlie our imagined identities. The myths of Marjayoun and its founding often revolve around accounts of perpetual flight. Tales of war and exodus, told and retold over eighteen centuries, they often make mention of a dam that once enriched a desert and, as the tellers continue their sagas, of three birds, one of which beckoned a frightened group of wanderers toward what would become my family's home.*

This fable of origin starts in the rugged climes of Yemen, in the land of the Queen of Sheba, where the frankincense and spice routes of antiquity snaked through Arabia and Abyssinia. It was here that the famed Ma'arib Dam channeled water to the parched landscape, but its occasional breaches also brought — nearly two millennia ago — what, through tellings and retellings, has become a near-apocalyptic conflagration that drove Marjayoun's ancestors to the rich fields of the Houran. There, and in other locales, their desert princes and warriors became known as Bani Ghassan, the children of Ghassan. According to legend, they became imperial mercenaries, allies of Rome, guardians of trade routes, a source of troops, and a client state in wars with Persia. The name Rome, or Rum, still denotes their Orthodox Christian descendants in the empire's former domains.

More exoduses would follow. Among these tales of departure, now much less spoken and repeated, one began with an anxious coterie of subjects seeking sage advice from their leader: Should they flee the strife surrounding them? To solve the problem, the ruler, Abu Rajeh, brought out three birds. He plucked the feathers of the first before cutting the wings of the second. The third he left alone. Abu Rajeh then sat back as his subjects waited, deliberating over their leader's message. At last they hit upon what they believed it was: The bird with wings can travel as far away as it wants. So they did. They went to the place that became Marjayoun.

Before the arrival of Abu Jean, I had spent many hours in Marjayoun wandering about, reacquainting myself, discovering, absorbing beauties, documenting what was no more. The sights became familiar — thorny rose bushes swirling and madly unkempt, red tiles fallen and crushed, and a once-smashing curl of blue, painted, it seemed, to draw the attention of passersby to some handsomely crafted wooden shutters. Only a few of the traditional lemon trees, which in the not so distant past were planted in almost every garden, marked the path I usually took to the Saha. The bitter juice was once meant to ward off the evil eye.

Marjayoun is set on a plateau of muted and melded grays, browns, and greens, blended in harmony with the land's past. From high points beyond Isber's house, its surroundings can be surveyed. Beyond the town's entrance is the Hula Valley, in present-day Israel, where the finer families once kept prosperous estates. To the west of the town, over a ridge, the Litani River flows, its waters tucked beneath oaks, eucalyptus, and pines. On the other side are Mount Hermon and its peaks, which serve as borders of Israel and Syria. Beyond it are the Golan Heights and Quneitra, an almost abandoned city since its Israeli occupiers ravaged it as they withdrew in 1974. Farther is the Houran, the expansive plain of the Syrian hinterland, where Isber Samara traded for years in his quest for wealth and reputation.

Facing Marjayoun, on the mountain range's western escarpment, is Wadi al-Taym and the Arqoub, along with towns that, for centuries, formed a diverse but integrated tapestry of Christians, Muslims, and Druze along its valleys, slopes, and foothills. Marjayoun — historically the largest of them — has one Sunni mosque and a church for each Christian sect: Greek Orthodox, the largest; and Maronite Catholic, Greek Catholic, and Protestant. Next to Marjayoun is Dibin, a mainly Shiite village, as are nearby Blatt and Khiam, across the verdant valley. Between these two tiny towns is Ibl al-Saqi, a village of Druze and Christians, famous for its olives and vineyards. Kfar Shoba, climbing Mount Hermon, is Sunni, as is Kfar Hammam and Shebaa. Rashaya al-Fukhar, renowned for its colorful pottery, is Orthodox and Maronite. Farther away is Hasbaya, with its Druze, Muslim, and Christian inhabitants, their intersections helping blur distinctions.

The stones of the hills around Marjayoun are a pale gray that blends into the ascending terraces. The land itself is furrowed with creases and wrinkles plunging downward like the lines across Mount Hermon. The roads meld with rivulets, which merge with ravines, which cascade into gullies. Nothing is jagged. Everything is rounded, even the rocks that thrust themselves up from the earth. Perhaps most belligerent is the wind, rarely subsiding.

There are not many young trees around Marjayoun these days, but the older ones, majestic and tall, stand like sentries on the hillsides. Pines litter the ground with needles and cones. Along the roads are groves of olive trees, deferential, mimicking the landscape. The trees resemble sculptures, their trunks wizened remnants of time compressed. A fig tree across the street from my apartment appeared to be big and lush, even tropical. It swaggered almost imperiously as I watched a young girl standing in the street under its canopy of rustling leaves in the midmorning quiet. With a stick she pulled a branch down to pick a fig. Her movements were unhurried and deft.

Gone were the pickup trucks that often rumbled through the town, their bullhorn-style loudspeakers mounted on top. *Aluminum, iron, car batteries for sale.* There were no sounds but the same wind, caressing the trees.

Two days after our financial negotiations, Abu Jean called at 9 A.M. for me to pick him up. We headed to the Saha, where we met Faez, a thirty-year-old Syrian from Daraa who had agreed to work on the house. Wearing a floppy blue hat, Faez had a quiet, gentle disposition. *Adami* (polite) they would call him in Marjayoun. Along with him, we brought a green wheelbarrow, two claw hammers, two shovels with the handles detached, a chisel, and the biggest sledgehammer I had yet to see in my thirty-nine years.

At the house, Abu Jean led Faez to the vaulted stone room I had nicknamed "the Cave," then the kitchen, then the bedroom.

"Use your head and figure out how much it's going to cost," Abu Jean told him.

Precisely what Abu Jean had done days before, Faez did, repeating the gestures verbatim and nodding — okay, okay, *inshallah, inshallah* — as Abu Jean kept at it. "Think what you need, then tell me." Faez knew the game, and he wouldn't be trapped. The rule: no commitment until the last possible moment.

"How much do you want, and tell me," Abu Jean said yet again. "One hundred thousand lira?" Faez laughed. Even I thought the offer was a little ridiculous.

Finally, Faez offered his number. "Five hundred thousand lira."

Abu Jean acted startled, falling back a dramatic step. "What's with five hundred thousand?" he asked. Mothers would be insulted, saints blasphemed by this figure. Abu Jean went to the wall, pulling off a piece of plaster with his hand to demonstrate how easy the work was. "You don't even have to swing a hammer!"

"It's not all like that," Faez pleaded. "The rest is concrete." Faez thought for a moment. "Take off fifty thousand," he said. "Four hundred fifty thousand is good?"

"That's a lot, *harram*," Abu Jean answered.

Faez was looking to close the deal.

"What's the last price you want?" he asked Abu Jean.

Abu Jean was silent for a few moments. "Three-fifty," he said. Then he actually winked at me in front of Faez.

Unbelievably, out of either satisfaction or exhaustion, Faez agreed.

As we were walking out, Abu Jean, a Christian, said something I hadn't heard since my mother used to say it, before we would depart anywhere in the car.

"*Itakalna ala Allah.*" It's in God's hands.

"*Ala Allah,*" Faez, a Muslim, answered, the simplicity of the response in itself beautiful.

A few minutes later, I dropped Abu Jean off in the market. He turned to me as he got out of the car. "Phone Hikmat and tell him not to worry about a thing."

In these first weeks, I often turned to my friends for advice. Dr. Khairalla Mady was a man who was truly respected. Kalim Salameh, an aging resident who had served for a remarkable fifty-five years as *mukhtar,* a kind of mayor, stared at the floor as he spoke about him. "He wouldn't take a penny from the poor," he told me. Kalim then echoed the words that I had heard many times from Hikmat, and even more so from Abu Jean. "The town," Hikmat would tell me, "no longer has men like that."

A patient, he said, would come to Dr. Khairalla with six eggs as payment. Others would bring cheese, or a creamy yogurt known as *labneh.* "So as to not hurt their feelings, he would accept them," Hikmat recalled. I eventually tracked down Dr. Khairalla's phone number. It was an awkward introduction; he was unaware of me. Yet he volunteered to pick me up, and a few minutes later arrived at the apartment. Hardly any conversation could begin in Marjayoun without an exploration of origins: The family would be recalled and characterized, with its labyrinth of relationships carefully delineated. Abdullah — the name of my grandfather on the Shadid side — gave the good doctor little to go on. Abdullah had departed Marjayoun too young to leave a legacy in the town, and even after all these weeks, I had yet to learn where his house had been. But when I mentioned his father, Ayyash Shadid, Dr. Khairalla gave a nod of recognition.

I knew less about the Shadids than about the Samaras, the family of my grandmother, so Dr. Khairalla took me to the old Shadid neighborhood. There — he pointed across an undulating street — are the Dahers. Over there, around the corner, are the Jabaras. Next to them are the Eids. And here, he told me matter-of-factly, is your grandfather Shadid's house.

It was quite a contrast to Isber's impressive villa — a simple, two-story concrete affair, next to the Serail, the old headquarters of the Ottoman government. This locale had none of the formidable stone of Bayt Samara; there was neither wealth on display nor arches attesting to taste. This neighborhood lacked grandeur and, in fact, spoke more of poverty, where flat roofs were once rolled with a stone in winter to keep the rain from pouring through heaving mud.

Dr. Khairalla lived down the street, and once inside, we sat and had coffee with his wife. Our conversation unfolded in four languages — English, French, Bulgarian, and Arabic. The doctor, trained as a urologist in Eastern Europe, had, in 1985, become director of Marjayoun Hospital, which had been founded in 1960 by a distant cousin of mine, Dr. Michael Shadid, who had raised $30,273 for the project from 203 donors in Oklahoma, Kansas, and elsewhere. (My grandmother was one of them.) A black plaque inscribed in gold still hangs at its entrance: "The Haramoon Charity Hospital conceived by Dr. Michael Abraham Shadid and realized through the contributions of the emigrants from Wadi al-Taym and Marjayoun."

Dr. Khairalla had managed the hospital for sixteen years, through a difficult time for Marjayoun and Lebanon, years that spanned war, isolation, and the Israeli occupation. It is a testament to his administration that today residents look at his tenure, hardships aside, as the facility's golden age.

Dr. Khairalla, I would learn, personified Marjayoun as it was, before war had taxed its charity, before change had disturbed what had once been its proud conduct. Now sixty-five years old, Dr. Khairalla no longer worked at the hospital but remained distinguished, with gray hair and glasses suggesting intellectual pursuits. Age had not expanded his lean build or obscured the line of the chiseled jaw framing his thin face. Yet as we talked until sunset, his words had begun to dwindle, and I realized he was tiring. I had heard he was very ill, news confirmed by his obvious fatigue. A little gingerly, with hesitant steps, he walked me to the street. I looked at his garden as we left. Some of the fruit trees were desiccated, others looked unpruned and unmanicured, abandoned to their fate. Patches of the terraces were overgrown with weeds, tangling in what he had planted. Here and there, stones had given way in the retaining walls that once held them up. I thought of a proverb Hikmat had told me: "The blessing goes with its owner."

Cancer is rarely spoken about in Marjayoun; the name itself is a harbinger of bad luck. To mention it is a grim omen, as if inviting its onset. "That disease, may God save us from it," people might say instead. Hikmat called it *al-marad al-khabeeth,* the wicked disease. "Most people don't talk about it," he told me. "They keep it confidential." He shrugged as he explained the doctor's lack of acknowledgment of his illness. "He'll tell you he's sick, but he won't tell you what it is." Dr. Khairalla had cancer of the prostate, and though he didn't mention it, I knew it was serious.

The dignity of Dr. Khairalla lingered with me for days as I turned to books he loaned me, really no more than folders of photocopied papers passed on from one hopeful friend or relative to another. Through the blurred words was the history of my adopted community that was unfolding before me, arrayed as it was in celebrated stories and vestiges of stone. In these weeks my spirit ebbed and flowed. I was not in a war zone, but I was not quite home. With little regret, I was enjoying the respite from work as a journalist, and all the dread that all those deadlines inspire, but I was missing my daughter Laila, just six years old. To her, I was an untethered voice on a cell phone, which in Marjayoun never worked all that well.

The modern incarnation of Marjayoun began five hundred years ago, and residents insist that at the turn of the nineteenth century, when Beirut remained a provincial capital, Marjayoun rivaled it in size. This statement is a reminder of the town's grandiosity, its tendency to inflate the past, a habit that one discovers in places more proud than vital. Yet looking back brings shame, or nostalgia no less common.

In comparison to the glories of the past, the reality of today in lands where war has never ended creates a kind of tristesse that underlies all and leads to further conflation of what has been. A more sober Ottoman report on the more ancient population, written with a sneer by two curmudgeonly Ottoman men before World War I, put the population in 1912 at 3,752 (2,195 Orthodox, 890 Catholics, 370 Protestants, and 297 Muslims).

Perhaps 800 people live in Marjayoun now, a shadow of the thousands in residence in its heyday, when nearly every morning, peasants near Ibl al-Saqi, Metulla, and across the Arqoub brought ghee, milk, and *labneh* to sell in the Saha, paved in black volcanic cobblestone. Most butchers slaughtered on Sunday, and Friday was market day, when vendors threw straw mats on the stone and hawked their goods. Crowds gathered to inhale the roasted nuts and tobacco burning in water pipes. The idle played backgammon near the shoe shiner; others gathered in chairs outside the barbershop, enjoying the breeze on freshly shaven and scented faces. A few sneaked drinks of arak in a store that, properly secluded, was a sort of speakeasy. As they did, they heard the vendors' cries, their accents diverse.

"*Bateekh, bateekh, bateekh . . . ala al-sikeen ya bateekh,*" one yelled. Watermelon, watermelon, watermelon . . . a watermelon ready for the knife!

The voluble merchant competed with others: "My cucumbers look like a baby's fingers!" "Tomatoes . . . red and green tomatoes . . . tomatoes from the valley!" "Radish is good for you but makes you fart!"

Their muffled cries were heard inside the store where a Christian vendor filled prescriptions written by a Shiite sheikh in neighboring Dibin for patients of all faiths. No matter what the disease or complaint, the sheikh, a savvy assessor of a placebo's worth, prescribed a variation of two remedies: either a compound of tamarind, anise, and fennel, or one made of honey, eggs, cinnamon, nutmeg, cloves, ginger, liquid styrax, benzoin of Java, and pepper. (Prescribed them, that is, if his elaborate incantation failed.)

In the Saha, children, unaware of the class consciousness that ruled Marjayoun's stratified society, played together with pebbles, or balls, or the bones of goats or sheep, which they begged from the butchers in town. Slingshots were just as popular, and were the scourge of birds whose migration patterns brought them through the Hula Valley and along the Great Rift Valley spanning Africa and Asia. On Sunday, they flocked to a small pool where youngsters played. As it happened, the pool was the creation of an enterprising resident who had built it by damming a stream running from one of the springs.

The young Isber Samara loved to splash around here, especially on summer's humid days. Perhaps it was where he first glimpsed the lives of others more fortunate than he, children with more than old slingshots to play with, fathers with clothing finer than he had seen before, gentlemen who spoke of history as if they had been granted seats to observe its unfolding.

Isber Samara, born in 1873, was the son of Suleiman Samara, who was the son of Ibrahim, who was the son of Mikhail, who was born in 1770 to a man named Samara Samara, who had settled in Marjayoun after coming of age in the Houran. Later, it was to that steppe that Isber and his brothers — Faris, Rashid, and Said — returned as young men to toil and push themselves toward the prosperity that was the object of all their hopes. The brothers started out as landlords for sharecroppers (which netted them eighty percent of any wheat harvest); herders of sheep, goats, and camels; and moneylenders who forced clients to guarantee their debts in land. They were also merchants, engaged in a brisk trade in jibneh baida, a kind of cheese, labneh, and samneh, or ghee, wrapped in sheepskin and carried on their horses.

Ambitious young men, the Samara brothers were aggressive, conscious of pennies. Isber, in fact, was reputed to occasionally put his foot on the scale as the brothers weighed the ghee and wheat that they traded with the Bedouins. The Samaras spoke their own dialect, Badraji, a distinctive vocabulary that cloaked their intentions before those Bedouin counterparts.

The desert ways, customs, and attitudes of the Houran felt familiar to them; of course, it was part of their legacy. The Samaras lived by standards disavowed by Marjayoun's gentlemen, forms and practices that marked existence in the desert and the Houran. The Samaras grew up speaking an Arabic more Bedouin than the pronounced Lebanese dialect spoken by the older families, like the Shadids in Marjayoun. They shared those Bedouins' words: "South" was never janoub, *but rather* ibli, *the direction of Mecca in a nomad's mind.*

They shared as well the Bedouins' keen sense of hospitality and honor, those conventions that had regulated existence for millennia in the desert, where survival might depend on a neighbor's goodwill. Like those nomads, they had contempt for the peasants whom they catered to, and for the blacksmiths, grocers, and other shopkeepers, with their regular hours and routines, whom they relied on. Isber and his brothers embraced a desert code versatile but unambiguous, the product of its creators' keen understanding of life's dangerous caprice.

At the center of the Saha in the days of prosperity was the *manara*, an Ottoman-era lamppost that, in popular lore, served at least once as an imperial gallows. It was built in 1908 by Hamdi Bek, a middling Ottoman potentate called the *qaimaqam*, who also installed the town's first gas lamp. Around the square he embellished was the mosque, the provincial headquarters, and perhaps eighty stores drawing their distant customers — shops for hardware and groceries, fruit and vegetable sellers, sweetshops, bakeries, barbers, and butchers. The marketplace was loud with confusing greetings, shouts of dissent in disparate dialects, outrageous obscenities alternating with clattering backgammon stones, and the cries of vendors' offerings.

The most famous of the Saha's buildings was undoubtedly the restaurant owned by the Akkawis, a Sunni family. Pots and flowers surrounded an atrium fountain around which tables and chairs ascended in tiers. The Akkawis' fresh fruit sorbets were famous; the ice to make them was brought daily from the highest peaks of Mount Hermon. On Good Friday, prayers would play from the speakers of the Muslim-owned restaurant. A young Hana Shadid, extolling God, would repay the favor.

Every year, diners in the atrium would see Hana, a Greek Orthodox Christian, climb the minaret with dignity, turn to Mecca, and begin to recite the Muslim call to prayer, touching each word with care.

Many years later, in Oklahoma, relatives would smile at the mention of this scene. They would recall Hana, whose singing voice was not unremarkable, yet more meaningful was the statement that he and the town were making together. For more than a hundred years Hana's call would be remembered by Marjayounis, separated from families and scattered across the world, when they encountered each other. In other times, less peaceful, they would marvel at the Muslims' acceptance of a Christian man addressing their God as their intermediary.

In a grocery store on the outskirts of Oklahoma City, my aunts — Najiba and Nabeeha — would recall the faces watching, unified by what Hana was saying for the town, which existed as it did because its people had learned tolerance. Here was Hana Shadid seeking God for everyone in the Saha, for everyone in the town. Here was a statement that would be heard around the world for decades. Here was Hana Shadid seeking God for all kinds, for all men and women. Here was a gesture that would say, "This is Marjayoun."

Michael Shadid, the twelfth child of a distant cousin, had been raised by a mother who could neither read nor write. In Marjayoun — a place he would one day chasten for its "little regard for vital statistics as for sanitation" — nine of his older brothers and sisters died in childhood, too sick for saving by a doctor whose only resources, Michael remembered, were "mustard poultices, and quinine, and calomel." His father had left his mother, Khishfeh, a one-room house, two mules, and a modest sum of money, which would sustain them for years. Determined to make it last longer, she appealed to God, burned incense before icons, and knelt to pray every evening. To earn more, she baked, washed clothes, scrubbed houses. Always, when shopping for clothing or food, she remained the model of parsimony. Her three children, always without shoes, ate a pastry once a year, at Easter. At mealtime, they had little more than rice, cracked wheat, dried figs, olives, bread, or honey.

The more fortunate in Marjayoun during these earlier days had land, a profession, or a trade; no one wanted to look the part of a peasant. Tilling the soil was looked down on. Pride in their education, Marjayounis called it; arrogance, said their neighbors. Men walking in the street kept pens in their pockets and newspapers under their arms. (When Hajj Assad al-Basit, from neighboring Rashaya al-Fukhar, met the French high commissioner in Marjayoun after World War I, he introduced himself as a subscriber of the newspaper *Al-Ahrar*.) Literacy and learning were things to be proud of and displayed. The town had three schools, Greek Orthodox, Roman Catholic, and Protestant — the last of which was considered by the more traditional to be heretical. The Russian Empire, as the most powerful Orthodox state, supported the town's Orthodox school, the Madrasa Moscawiya, literally the Muscovite School, and its four hundred or so students. For years, some of the elderly could still muster a few words of the Russian taught there.

Four newspapers were once published here, each with its own bent: *Sada al-Janoub* (*Echo of the South*) leaned toward the liberal, *Al-Marj* (named after the town) was inclined to a vision of a Greater Syria, *Al-Qalam al-Sareeh* (*The Forthright Pen*) was vaguely Arab nationalist, and *Al-Nahda* (*The Renaissance*) adhered to a narrower definition of nationalism. Societies dedicated to that *nahda* sprung up in a confident town that deemed itself cosmopolitan, a capital in its own right of the hinterland beyond.

Marjayoun was divided into two neighborhoods, bisected by the main road that came to be known simply as the Boulevard. In the west, climbing the plateau, was Hayy al-Qalaa, the neighborhood of the citadel; below was Hayy al-Serail, the neighborhood of the headquarters. The wealthiest families here — the Farhas, Barakats, and Gholmias — had homes of stone and red tile equal to those found anywhere in historic Syria.

To houses of the rich and poor in Marjayoun, priests once visited twice a year, at Eastertime and again on the day commemorating the baptism of Jesus in the Jordan River. They carried pails of water already blessed and sprinkled it on the walls of the houses. Unpaid, they relied on charity from those blessings, as well as the fees they expected for performing marriages and baptisms, conducting funerals, and locating brides or grooms.

Then as now, religion was everywhere, beneath the cadence of life.

Oaths were taken on the life of the Virgin, the Messiah, or the cross. Thanks offered to the Almighty produced strength for labor, extended life, and cleared the way for any traveler. For unfinished vendettas with the dead, God was implored to bury them yet deeper.

But God was not the only authority. The Serail, the stately building of columns and arches built as the local outpost of the Ottoman Empire, was the symbol of power and order. It was here, in an earlier time, that a bureaucrat, perhaps considering lunchtime and a glass of arak, mistook the year of Isber Samara's birth and drafted him, at age forty-three, into the Ottoman army. Three years later, Isber would be sentenced to death.

The drafting of Isber Samara was almost certainly a clerical error, but more pressing duties than paperwork preoccupied Ottoman officials allied with Germany in World War I. Short-handed and suffering, the Ottomans drafted young men from towns that had for centuries been spared from military service. Scores of eligible males fled for America or Brazil, spending whatever they had to board the last Italian steamers departing Beirut before the port was closed by blockade in 1915. Those who remained hid, anywhere. One man was said to have walked the streets disguised as a woman.

Isber's birth certificate stated that he was ten years younger than he really was, but in his opinion, the errors of clerks did not merit response. The Ottomans, it seems, were of a different mind. Upon his return to Marjayoun, where the Turks kept a small garrison, he was arrested and sentenced to death. For weeks he awaited execution in Marjeh Square, the Ottoman administrative and commercial center in Damascus, which was known for its glistening bronze colonnade, erected nearly ten years earlier by Sultan Abdelhamid II.

Bahija was told little about her husband's imprisonment, but she surmised all. She had six children to care for and, unlike her husband, did not see danger as something to be courted. She did not imagine the events of a day as things that bowed to her. As Bahija practiced patience, Rashid — the youngest of the remaining Samara brothers — who had none, pleaded with Faris to act. As the date of execution neared, Faris — sixteen years Isber's senior and as boisterous as his brother was taciturn — gathered all the family's gold, traveled to Damascus, and began disbursing bribes, to no apparent effect. As time dragged on, Faris feared the worst, but before Isber was to perish, his brother succeeded, though the payoffs had taken nearly all they had saved over the years.

Cheating the executioner seemed to have changed Isber, or at least furthered his sense of himself as special, even chosen. He had confidence. More and more, Isber seemed to believe that life would follow in the direction he pointed. I am not unfamiliar with the relief of cheating death. But one victory does not, obviously, cancel the fact that many others were not so lucky. Some are haunted by being spared, never feeling deserving of their discrimination. Others see their luck as an act of affirmation, the bow of destiny to their far from ordinary distinctions. Isber was one of those. I hope that I was not, but I was younger when I was lucky enough to survive a bullet. I did not always understand or distinguish things that I hope I would today. I am Isber's great-grandson, after all. So I have learned.

· 4 ·

OUR LAST GENTLEMAN

I HAD MET SHIBIL before the war in 2006. Twenty years my senior, he had dark hair cut short like a teenager's and a swath of mustache trimmed with a precision he seemed to direct at almost nothing else. His eyes often flashed heightened surprise, and his ever-expanding gut seemed to grow in proportion to the angst it caused him. We had in common Oklahoma City, where I was born and where Shibil, in the 1970s, going by the name Sam, worked, occasionally studied, and regularly consumed a vast amount of drugs. All these years later, marijuana was still his staple, and he remained its proponent. He had the bemused detachment of a perpetual stoner; nothing was ever wrong or urgent.

On the day I recall, I was taking a break from work on the house and found Shibil at home, standing on his balcony in boxers, no shirt, a scotch in progress. His Persian carpet was draped over the railing, after a rare cleaning.

"*Keef sahtak? Mashi al-hal?*" he asked in a hearty voice.

I walked in, and Shibil poured me a drink.

"*Kasak,*" I said to him.

"*Sahtein, sahtein,*" he replied.

Then he added in English, "Cheers, man."

Shibil was always hospitable, far more so than others here. Like those of earlier generations, he took pride in his manners, which were strictly old family. Marjayounis, according to one of Shibil's friends, always insist, "Next time you're here, come have lunch." Yet when the next visit arrives, they do not extend an actual invitation, but repeat the previous suggestion. "Next time . . ." Some lacked even that grudging formality. I had already come to dread the cackling voice of an elderly neighbor echoing off the pavement outside the place I rented, as she shouted at her husband or daughters or both. She would traipse through the street, her hair uncombed. Once she had ventured over to pick flowers on my porch, meeting my look of surprise with a stare that I thought suggested satisfaction. Shibil's hospitality, conversely, was unstinting and gracious. After my first scotch, he brought a glass of ice and set the bottle of Grant's on the floor. "*Hutt whiskey!*" he bellowed. Pour whiskey!

Next to the glass were pickles and an array of snacks — cashews, pumpkin seeds, peanuts, and pistachios. Things got a little awkward when Shibil, unexpectedly and casually, tuned the TV to Spice Premium, a porn station. To the backdrop of an especially vigorous sexual interlude, we talked cooking — he suggested wrapping cucumbers in grape leaves to make them pickle faster — then eased into politics.

He declared that 1973 was the last good year in Lebanon whose wars and crises had apparently exhausted him. He seemed to grow more fatigued as he spoke of 2006. The war that had brought me here had ended, but its aftermath embroiled virtually all in Lebanon as the government tried to contain Hezbollah. Israel, of course, was watching carefully in the background.

"A big surprise," Hezbollah's leader promised Israel, if it dared attack again. No one seemed reassured.

"Fuck this shitty country," Shibil offered.

His health was a matter of unceasing concern. As our visit continued, Shibil began a complete report on his upset stomach. Then he advanced with accumulating detail to an operation he'd had to seal a fissure in his rectum, then on to a bout with diverticulitis. He had almost died. Attempting to lift the shirt stretched taut across his ample belly, he tried to show me his scars. Soon, though, the conversation returned to the house. He would not let up, again demanding that I not rebuild.

"People in Jedeida don't think you're crazy, Anthony. They think you should be in an insane asylum," he said. "They think you should be locked up." Then he added, "Don't get me wrong, but you're an American.

"Moussa Barakat!" Shibil suddenly exclaimed, for no apparent reason, as he began the finger wave with which he signaled announcements of useful instruction. Then he got to the point: Fired by pride and history, Moussa had wanted to renovate his house, and he tried to strike a deal with his cousins. They could take the farms and citrus orchards the family owned in Jordan, he said, and he would assume ownership of the house in Marjayoun. If there was a difference in cost, they could settle up with each other. After he spent $150,000 on the renovation, his cousins said the deal was no longer in force, despite its having been committed to paper. At first, Shibil said, Moussa couldn't even enter the house, so chagrined was he by the pettiness. Now, he doesn't like to go there out of anger at what his cousins did.

"Listen, motherfucker, be careful!" he said. "If I'm wrong, I'll cut my dick off and eat it."

I laughed. "I agree with you, Shibil. It's a risk, and God knows what's going to happen. But imagine, I can bring back something that was lost."

"It's better than fucking Kunta Kinte," he said. Nothing would sway him about the house, though.

"The people in Jedeida are down-to-earth. They're very kind and nice. Until it comes to property." When he repeated the proverb "A sliver of land can wipe out its people," I felt suddenly at peace, tranquilized by his sentiments and secondhand smoke.

"Keep it as a reminder," he said, "that's all. I'm not joking with you. They'll come up from nowhere like plants. That family. Like those weeds that grow. You won't know where they came from."

He looked fearsome as his jowls shook. "That's the mentality of Jedeida."

Wheelbarrow after wheelbarrow piled with rubble left my grandmother's house daily. Abu Jean wielded a crowbar, loosening the cinderblocks of a wall much older than I am. With each exertion, dust — gray and brown — trickled down in rivulets. As the wall shuddered, Faez swung a sledgehammer into what was once a bedroom downstairs. Another worker, Malek — surlier, reminiscent of a street tough in Cairo or a petulant militiaman in Baghdad — stood atop a barrel, chipping away at the plaster in the Cave. He methodically removed each piece, as Faez, more diligent, fluttered from wall to wall in a frenzy of demolition. Their faces were covered in dust, the same color as the stone.

Four times, Faez struck as I stood there, before the plaster creaked. Then the cinderblocks shifted and buckled. I was torn between exhilaration and dread inspired by watching the work. I waited for the ceiling to collapse, the house to come tumbling down around us. For a moment, for my own safety, I thought about leaving the room.

"Just one more cinderblock," Abu Jean told me, reassuring me with a wink — that was all he needed before his wall would give and crumble. It did, spilling across the ground.

There was meaning to the destruction, an elegance of movement as the house hurtled toward its end and a new beginning. Clouds of dust billowed out of the door, caught in a breeze; in moments, they evaporated. Wind blew through the house, raising a haze that caught the late-afternoon sun. Cigarette smoke danced with dust suspended in the air. Sparks flew as a pickax collided with stone. The lights arced, then were extinguished in dust. The sounds were relentless, like a drum in martial cadence.

The days passed, and more followed. Rooms that squatters had called their own disappeared, as did the marble kitchen sink, along with the bathroom fixtures, demolished or dismantled. So did the artifacts left behind whose origins I never knew. With each trip of the wheelbarrow, with each swing of the hammer, another decade of the house went with it. It took that much force, brute force, in the midst of what seemed a whirlwind of destruction, to remove what the house had suffered, to detach what stood between now and then. And as each decade passed before me, the house unveiling itself, I began to see into the past.

Perhaps because I had been so long discontented with the world around me, I increasingly turned my attention to Isber's world, which, while simpler, was no less tumultuous than my own. Isber built this house as his citadel, so sturdy and lasting it seemed to rise from the bedrock itself. It would compensate for his rough edges and offer his family shelter and a future, an inheritance. My family wasn't here. They had shown little interest in my project. On those occasions when I spoke to Laila, she often asked me what I was doing so far away. "Rebuilding our home," I told her, but understandably, given her age, she failed to appreciate that this absence was, oddly enough, my attempt to make amends for all the others. If I could: Our lost home in America could not, after all, be restored, and perhaps this project offered only futile compensation, satisfying, finally, only to myself. When the doubting part of me took over, I turned my attention to the work. For some reason, maybe for my family in Oklahoma, growing older and frailer, maybe for something else, I had to go on. It was *bayt* and the desire to resurrect what once stood for something.

We all wondered what might turn up as we returned the house to the essentials that Isber Samara had envisioned. Weeks, even days before, it had felt modest, cramped, and dark, some rooms unbearably claustrophobic. In its destruction, the house, liberated, revealed its origins — stone, hewn a century before. Vaults were hitched to the bedrock that served as the foundation. In chaotic geometry, smaller stones climbed over each other in the rugged, disordered perfection of the Cave's arcade. Greater stones formed wall after wall, perfectly laid by masons from the north of Lebanon. Their angles were designed in strength, their beauty incidental. More stones constructed the two arches, side by side, once the stately entrance to this house that my great-grandfather built. For years they were filled in with stones underneath, their curves becoming the lineaments of a wall. To add insult, they were then buried in concrete. Now they were restored, portals to another time, the era when the Ottomans fell and when Isber Samara became not quite a gentleman.

The harvest of 1918 was good. Druze, Muslim, and Christian alike sold their products to the warring armies in World War I, Ottoman and British, trying to feed hundreds of thousands of soldiers. Well stocked with gold, the British and their Arab allies bought grain from the Druze. Of their six hundred bushels, Isber and his brothers sold half to the Turks, but hoped to make a bigger profit on the rest. Finally, Isber agreed to sell the remaining stock to the Turks, but refused to take their paper money. Nor would he accept the Ottoman gold pounds. Isber, who had come to believe what no one ever had — that the Ottoman Empire could fall — demanded the British gold sterling in payment. The Turks had no other choice. The gold went to the Samaras.

On the day of the deal that transformed the Samaras' lives, there was no thought of what might lie ahead, as the children of the brothers, later separated by thousands of miles, saw a sight they would never forget: There they would be, always in memory, their fathers, running like boys toward them, advancing toward home, free from care. One of the children, long grown, would recall with a bittersweet expression those gold pieces piled as high as a hill in the family's sitting room, glowing in the light of dusk.

One morning, taking some time off-site, I met Shibil at around 10 A.M. and we headed to the Friday market in Marjayoun. The village market is an old tradition in the countryside, a legacy of Bedouin culture, and many towns around Marjayoun had them. Qlayaa's was on Sunday. Monday was Nabatiyeh's. The most famous was Suq al-Khan, on Tuesday, near Kawkaba. They were really no more than amalgamations of hastily erected tents and rickety stalls, where everything from scarves and screwdrivers to corkscrews and pirated CDs were hawked. "Beautiful prices!" vendors shouted, to no one in particular. However pronounced the tension, and even in times of war, Shiite butchers hung out their meat, willing to cut a slice and grill it, and Druze farmers, in their white knit caps and baggy pants, kept coming to sell pickled wild cucumbers and cauliflower.

Shibil sometimes called himself Oklahoman, but he was really, inexorably, a son of this town, a belief confirmed as I watched him cringe when a black goat crossed our path. Like the evil eye, it was an omen, and omens mattered. In winter, he would never walk outside without splashing cold water on his face. His superstitions continually announced themselves. "Beware of split teeth and blue eyes," he warned me as he scanned the market crowd, deadly serious. "Small foreheads, too."

As we walked, he regaled me with stories of the town's citizens. There was Farid Maalouf. His fingers, Shibil told me, had four joints rather than three. With his massive hand, Farid once knocked a horse to the ground. Shibil's eyes grew wide in recognition: Nabih Abu Kassem, he remembered; there was a man. "God have mercy on his soul," Shibil said, genuine sorrow in his voice. A decade or so ago, Nabih and his friends were drinking in the morning when they got hungry. Without missing a beat, Nabih got up from his chair, grabbed a knife, and ambled outside. He found a sheep and, from the flap of flesh behind the animal's buttocks, he sliced off a marble-colored slab. The sheep's cries followed him as he returned inside. He skinned the slab of fat, sliced it, and salted it, sharing with his friends, who kept to their drinking, impervious to the wails of the wounded sheep.

He was no less steeped in the culture and traditions of Marjayoun, which despite the town's decline retained some of the features of Isber's day: Slights here became grievances, and grievances became grudges lasting months, years, sometimes decades. Shibil's latest was directed at the local grocer, Saad Barakat. It had begun when Shibil walked into Barakat's establishment and sampled two almonds.

"Should I open a beer for you, too?" the grocer asked. Then Shibil took a Kleenex from the counter and wiped the top of a soft drink can he had just bought. "Am I Caritas?" Barakat asked.

"What are these shit people?" Shibil asked as we walked.

Hospitality was a memory, he insisted. Nothing was being honored anymore. Only the elderly, Shibil groused, still went by the old tradition of taking the name of their oldest children: Abu Ghassan (Father of Ghassan), Abu Jean (Father of Jean), and so on. The accent of the place was also disappearing. For centuries, Bedouin-inflected Arabic had been spoken here. But just as English was colonizing the world, the accent of Beirut was displacing Marjayoun's colorful, distinctive dialect. Shibil loathed it. Instead of *baarafsh* (I don't know), he declared, as we walked through the market, "they are saying *ma baarif.*" "Nothing" is now *ma fee* instead of *feesh*. "I don't want" had become *ma badee*, not *badeesh*. Soon, he suspected, we would start losing words that belonged specifically to the town, such as *kooz* (a small water pitcher) and *dashak* (a seat under which you can hide something).

To Shibil, it wasn't simply a loss of accent, not just a flattening of the diversity and integrity of the history bequeathed to the town. It was all part of the loss of identity and the loss of the influences that had made Marjayoun what it was. The desert dialect had arrived from the steppe of the Houran; the Palestinian inflections were brought by the tradespeople who had crossed the Hula Valley and beyond. In a way, Shibil suspected, the loss of the dialect was the loss of the last relics of the town's glory.

"The only time people arrive here is when they're dead," he said. His voice was tinged with helplessness, perhaps surrender. "They bring people here to bury them."

World War I ended soon after the Samara brothers closed their grain deal, the Ottoman Empire vanquished for the last time. In April 1919, Isber made his way home from the Houran as a blustery wind blew his baggy pants and camelhair cape. Spring had declared itself; the shaqaiq noaman, *with its ground-hugging, deep red petals, washed up against the soft whites of daisies weaving across rocky hillsides of an otherworldly gray.*

Toward the end of that valley Isber crossed was the Jewish town of Metulla, its land purchased by Baron Edmond de Rothschild in 1896.

The French and British victors of World War I were already imagining along Metulla's outskirts the border that would, in time, divide Lebanon and Israel, and even then the imperious French authority loomed over the distant Houran. The lands beyond Marjayoun were seething. In the spring of 1919, six months after the war's end, the landscape of the Samaras was still not yet Lebanon (it remained Syria, or bilad al-Sham to some), but had lost track of anything it was. The space left by the Ottoman rulers, once considered eternal, was showing itself deep and wide.

First to arrive was an army of Arab nationalists, one of the groups vying to take over parts of the Ottoman lands. It was led, in name, by an Arabian chieftain, Sherif Hussein; in reputation by T. E. Lawrence; and in action by Hussein's third and most popular son, Emir Feisal. Leery of French and British intentions, Feisal had instructed his men to declare an Arab government as soon as the Ottomans left and before the Europeans arrived. A proclamation announcing the new leadership reached the town, carried by 120 weary but resolute horsemen.

Mourad Gholmia, a former deputy in the Ottoman Assembly of 1909, a partisan of Feisal and a resident of Marjayoun, would recall scrambling with his compatriots the night before the proclamation to find fabric for the flag they would raise at the Serail. Gholmia himself proclaimed the government to the bewildered crowd that had gathered, a speech punctuated by the firing of twenty-one rounds.

"It was the first Arab flag to be raised in Beirut, Syria, Aleppo and Palestine," he later wrote proudly. For years, Gholmia would entertain anyone interested in hearing the story, though his moment of fame was almost as brief as Feisal's state which followed.

These were not likely to have been matters that concerned Isber. Not yet. Flush with wealth and heedless of the tumult, he was determined to take his place among the families that had looked down on his own. But even in the upper reaches of his own clan, his wealth had not changed his status. Distant cousins—haughty, and his only "significant" kin—refused to sell their stately villa to him, so he and his brothers decided to build their own houses next to each other, all drawing from an aesthetic more Roman than Greek. Before the war, they had purchased the land from the Gholmia family, one of Marjayoun's more prominent, in Hayy *al-Serail. Isber chose a plot of land above the villa he had coveted, near a spring called Kharrar, for a house that he hoped would last hundreds of years, just as the homes of the other prestigious families of Marjayoun had. He had begun the house years before. Back from the Houran, World War I over, he would finish his task.*

As Shibil, whose house was also, in more eccentric ways, his signature, talked in the Friday market, a tall man in sunglasses approached. Jutting his chin forward, he walked with purpose, as though posing for a photo shoot. Karim, as he liked to make clear, was a Shadid (though his family had adopted the name of his great-great-great-grandfather).

My cousin was alternately — as was immediately apparent — engaging and suffocating. (Shibil once said that Karim could speak for three or four hours with hardly a pause.) He had a bone to pick with me. It had all begun with an article I had written for the *Washington Post*. Sentimental, perhaps generously so, it celebrated olive trees, Mount Hermon, history, and family. "Lebanon, My Lebanon" was its title, admittedly grandiose. One sentence, penned as an afterthought, concerned the reputation of the Shadids: "Crazy, they would say, more so as they got older." I had thought little of it, especially after it was published. Karim, it appeared, had thought about little else.

"You were misinformed," he told me, his eyebrows arched in scorn.

He cast a glance at Shibil, then gestured at me to be silent. We would discuss the matter at lunch, alone. "As cousins."

As with many people here, Karim was reluctant to divulge his age. Fifties, he said vaguely, leaving it at that, though his hair was dyed the same deep black as my landlord's. Well shaven and manicured, he chose his subtle barbs with the same precision with which he pressed his clothes. His sunglasses recalled Anna Wintour's, and he tossed his head slightly but often. Karim, I have to say, was dramatic.

Like his siblings, Karim was an educated man, having studied at the International College, a prestigious high school in Beirut, then at the American University of Beirut, where he eventually received a master's degree in political science and economics. He earned his law degree, concurrently, at the Lebanese University. His French was good, as was his English, which bore the inflection of his time studying at the London School of Economics. He practiced as a lawyer, though work seemed intermittent.

Karim's story was of a sort not unfamiliar in the years of civil war. His father, an accountant and manager at the Iraqi Petroleum Company, died young, at fifty-one, of a heart attack in his sleep. Karim was just seventeen. Sadly, he recounted the fact that his father had built a new home in Marjayoun without ever sleeping a single night in it.

Karim went on to tell me of his sister's equally regrettable end. A doctor specializing in pediatric hematology, she had taught at the American University of Beirut, and died at thirty-four, on June 12, 1976, which he called "the worst day in the history of Beirut." There had been no electricity. Fighting between Syrians and Lebanese leftists allied with Palestinians, coalitions mercurial and agendas opportunistic, had paralyzed the streets. Another sister had been lost at fifty-four to breast cancer.

"We had too many of those evil eyes around us," he always said, by way of explaining why he occasionally burned incense in the house and garden. "I pray and I ask God Almighty to keep the evil eye and the envious eye away from us."

The next day, I saw a note stuffed in the door of my apartment, scrawled on tissue paper: "Toni: Passed by to see you but unfortunately you were not in. Karim."

An hour later, he was back, banging loudly on the door. There was no such thing as privacy in Marjayoun, I was learning, nor a call ahead. I had an omelet on the burner, and my expression was slightly harried. No matter. Karim expected to be invited in.

"I just want to stay for a few minutes," he said. That meant at least an hour.

"Christo Santo," he said as he entered the apartment, wearing a navy-blue T-shirt with a white neckband and jeans. An array of invective followed, punctuating the latest gossip, as he made his way into the house. "Son of a b," he said in a high voice. Then he spelled out b-i-t-c-h. Instead of *kiss umak,* Arabic for fuck you, he shouted, "Kiss me!"

We sat down at the table, and I ate my omelet, overly soaked in olive oil, in front of him. I offered him something to eat, but he declined. Then his interrogation continued as if it had never ended. How much did my stove cost? What about the water heater? What was wrong with it? The thermostat? (He could fix it.) Did I need olive oil? How much was the most recent plant that I bought for the house? Thirty-five thousand lira, I told him. He had heard that I paid thirty thousand, he told me.

I looked at him, shaking my head in disbelief.

"How did you know the price?" I asked.

"Hmm," Karim answered knowingly, then explained. He had heard that I was seen carrying a plant in the street, so first he narrowed down the possible stores, visiting each. Finding the right one, he asked how much the plant cost and how much I paid, making sure there was no discrepancy. He smiled at me, pleased with himself and his abilities.

A few days later, on a sunny morning, I sat with Hikmat and his wife, Amina, a Lebanese-American woman from Kentucky, who was pregnant with Hikmat's first child. The second of eight children, Hikmat had three brothers and four sisters. Of them, Hikmat was always bound closest to Marjayoun, where he and his father were born.

"What Marjayoun is, is memories," he told me. "It's a dead town but it still has memories." I understood what he meant. "For everybody it has memories," he said. "Nobody with those memories would sell his property."

He invited me for lunch, but when I asked, "Are you serious?" he bristled, marveling out loud at impolite Americans. "*Ffff*," he said. He turned to Amina. "I almost told him to fuck off.

"The food comes with you or without you," he told me.

A few minutes later, Shibil pulled up at the house in his rickety white Mercedes. He and Hikmat had been childhood friends but more recently had suffered through their differences. Shibil still visited anyway, and Hikmat welcomed him properly. We sat on Hikmat's balcony, and our host was expansive.

"Your life is like smoking a cigarette. You smoke it, you put it out, and it's over," he told me. "You have to be happy." Hikmat looked out at the nearby cemetery, a bad omen, I had learned, in Marjayoun. Even the tall, pencil-like cedars bordering it were said to speak of death. Hikmat wasn't fazed.

"I like sitting here," he said, as we ate kebab, balls of minced lamb, and chicken. "It's a reminder of that. There's an end. Be happy. Be honest. Believe in God. Be good."

Shibil nodded. "God give us peace," he said.

As we talked, the sun hit my face, and the call to prayer began from the Sunni mosque, faint in the predominantly Christian town but still resonant. Its lonesome, plaintive summons reached a crescendo, stopped, then began again with a declaration of the omnipotence of God. For a moment we listened, and Hikmat mentioned a thought I had heard occasionally over the years: There was a part of Islam in every Arab Christian. Shibil agreed. Whatever their beliefs, they acknowledged sharing a culture that bridged faiths, joined by a common notion of custom and tradition and all that it entailed — honor, hospitality, shame, pride, dignity, and a respect for God's power. For many Muslims and Christians there was even a common origin, a fabled beginning in faraway Yemen.

"If you cut me, you see Lebanon," Hikmat said, somewhat dramatically. "You see the Prophet Mohammed, you see Imam Ali, you see the cedars." He refilled our glasses with scotch and grabbed a piece of paper. Three sons had inherited seventeen camels from their father, Hikmat recounted to me, scribbling the numbers in a notebook. The oldest son was to receive half, the second son a third, and the youngest a ninth. The inheritance, though, was indivisible. The sons quarreled, then finally agreed to take their dispute to Imam Ali, the warrior, sage, and seventh-century caliph whom Shiites consider the divinely sanctioned successor to the Prophet. To solve their dispute, Imam Ali gave them one of his own camels, making eighteen. The oldest son then received nine, the second son six, and the youngest two — in all, seventeen camels. "Now give me my camel back," Imam Ali said.

"How can you not respect such a man?" Hikmat asked me.

Shibil shook his head. "Imam Ali was a great man," he said, and he quoted two lines from *Nahj al-Balagha* (*The Way of Eloquence*), the collection of Imam Ali's sayings, sermons, and speeches, which has served for centuries as Arabic's most exalted expression, much the way Cicero's speeches did for Latin.

"This happened fourteen centuries ago," Hikmat said.

The wind picked up, breaking the vestiges of summer heat. Shibil soon headed home, reluctant to miss his siesta. I made a gesture to leave as well, but Hikmat insisted I stay. The talk of history had recalled his father, George Farha, who in his lifetime had a reputation as a *zaim*, a word that can mean village luminary, strongman, or feudal lord. Hikmat still seemed determined to impress him. Seven years before, Hikmat had decided to return here, bringing to a close a peripatetic life that had taken him from Lebanon to Wichita, Kansas, to Saudi Arabia, to the St. James's Club, a resort on a secluded hundred-acre peninsula on the southeastern coast of Antigua. He rebuilt his house, exhausting what money he had saved. Here, he could be his father's successor. "I'd never shake your hand and ask you for something," Hikmat typically said, as his father always had.

As we sat at the table, Hikmat acted the part of the son of George Farha, scion of his family in Marjayoun, inheriting history, identity, and wisdom. In Arabic, the word "proverb" has none of its American connotations; here, hackneyed clichés become accumulated truths, and Hikmat recalled his father's favorite: "We have a proverb that says a proverb never lies." Sometimes the proverbs were playful, even witty. "Whoever marries my mother, I call him uncle," Hikmat said, or as he interpreted it: "If my mother wants to marry someone, what can I do? *Khalas*, enough. I call him uncle and make it easier on myself." Other times, proverbs were well intentioned but poorly executed. "I'd rather be a king in my country than a beggar in America," he told me.

Who wouldn't? I thought.

"Than the president of America," he said, eventually correcting himself. That was still mystifying, but better judgment suggested that I let it go.

Most often, a proverb offered a code of life, which was one of those same traditions Shibil had lamented only a few days before. "*Msayyar mish mkhayyar*," Hikmat told me. Loosely translated, it meant that life is ordained, not chosen. As the afternoon hours passed, Hikmat confided in me; I was an outsider, after all, and my distance seemed to allow him a vulnerable moment. "Whatever you do, God chooses for you. It's not up to you," he explained. "You don't make choices. You can't choose your direction."

Around this time, my cousin Karim had received the intelligence that my car had been seen at Hikmat's house. Karim had not been happy. Hikmat was conceited, he said sharply, and lacked his education. "He thinks he's a *zaim*, and he thinks his father is a *zaim*," Karim said. He stared at me a moment, then shook his head in contempt.

So began, in declarative fashion, my introduction to that other side of Marjayoun. Grudges between people gathered the resentments of nations defeated in war.

Both Karim and Hikmat agreed on what had divided them: Karim, it seems, had unsuccessfully run for office in the 2001 municipal election; for reasons too convoluted for me to comprehend, he blamed Hikmat for the loss.

"There will be a day when I return the gesture one thousand times," Karim said. There was hurt in his voice. "I'm a Christian. I don't bear a grudge, but he did me harm."

Hikmat related that for two years, Karim had refused to walk past his family's building. This seemed quintessentially Karim. He never shook hands with those who had slighted him. At best, depending on the severity of the previous insult, he would place his hand over his heart as a substitute for a shake. Other times, he would just nod — the greater the anger, the smaller the gesture and the quicker the movement.

Not much time had passed before I found out that matters had been made worse when Hikmat decided not to invite Karim to his wedding.

"Imagine," Karim remarked, still taken aback.

And then there was the matter of Hikmat and Shibil. "I feel sorry for him," Hikmat said. Although Hikmat was younger in age, he seemed to envision himself as the older brother to a wayward son. Upon the return of Shibil and a friend to Marjayoun after studying in Oklahoma, someone had declared, "They went as donkeys, and they came back as mules." Hikmat had been pleased by the remark.

By the time I arrived in Marjayoun, the bitterness had grown, perhaps beyond soothing. The latest slight seemed mundane. Shibil and a female guest were visiting Hikmat. Hikmat had momentarily left the room, and when he returned, he saw the guest squeeze Shibil's hand or pat him on the leg. (Hikmat didn't recall which.) He surmised that the two were speaking about him behind his back, though Shibil denied this.

"How dare they disrespect me in my own house?" Hikmat asked as he recounted the story. "You cannot help a person who's an enemy of himself."

In the days that followed, he would return again and again to the latest episode. *That woman had whispered something in Shibil's ear!*

"This son of a bitch. I come to your house and I talk behind your back? Fuck you. In my house? Small things sometimes cause big problems in life."

Sometimes, it seemed, drama nurtured the spirit.

· 5 ·

GOLD

THE EVENING SKIES grew clearer and crisper, offering vistas of Mount Hermon that I had yet to see.

I had plenty of time to assess the state of the mountains because there was no progress to measure at Isber's. After the satisfying paroxysm of destruction that initiated the project, the labor had returned to a village-like tempo. Meaning, very little was done. Days would go by, and I would find no one there. The house felt more desolate than ever, as the wind blew up eddies of dust in a pristine quiet. If I did find someone there, it was Abu Jean, with perhaps one other worker. Abu Jean would hire someone for the day, then stand and watch for hours, a Cedar cigarette dangling from his mouth, while the worker chipped away at the cement, plaster, and accumulated dirt that encased the house. *I'm in charge!* (Kind of.)

There were no other workers in town, Abu Jean insisted with papal certainty. We had yet to speak to a plumber. At the mention of an electrician, Abu Jean shrugged his shoulders. A neighbor suggested that at this pace it would take five years to finish. Grimly, I nodded in agreement. Abu Jean was aware of the frustrations; hardly a conversation went by without us discussing them. The answer was invariably the same.

In the first version, Abu Jean would put his thumb to his fingers, turn his hand upside down, and bob it. "*Ruq shwaya*," he would say. Take it easy. In the other version, he would look at me quizzically, then ask, "*Shou baddi aamel?*" What am I supposed to do?

So our conversations went, as days turned to weeks and the weeks dragged on through October. As in: Did we ask about an electrician in Hasbaya? Is the blacksmith coming today? The plumber? Did we tell the *maalim* for the roof shingles to come meet my cousin the engineer? What did the mason say about the stone wall behind the house? "*Ruq shwaya*," Abu Jean answered as the questions piled up. My protest followed — I suspected he didn't even hear me — as did his refrain: *Shou baddi aamel?*

"Abu Jean," I pleaded this time, exasperated, as we sat on the patio of his house, where his wife brought out a banana, grapes, and coffee. "I have no alternative. I have to finish this house, and I don't have a lot of time. If I need to pay more, I'll pay more. But we have to find people and we have to start working. *Now.*" I had practiced the speech, delivering it in my head a dozen times. Abu Jean's answer was no less practiced: "Do you want me to do it myself? What can I do? I'm doing everything I can."

And, of course, "*Shou baddi aamel?*"

Like a Greek chorus, his wife shouted out the same excuses.

"It's the *Eid!*" she yelled, a Muslim holiday.

Abu Jean looked at the grapes, untouched as we sat on his patio, and became indignant. "Why are you always in a hurry?" he asked. "Where do you have to go?" Abu Jean pushed the plate across the table and peeled the banana for me. More coffee followed, and this, a ten-minute visit, turned into an hour.

"I'm taking care of you better than I'd take care of my own children," he insisted.

And, at that, I surrendered. Isber Samara had envisioned something lasting in Hayy al-Serail, something worthy of the stockpile of gold that helped him buy his land. For a time, my vision was ambitious, as I conjured up pictures of cemento tile shaded in royal variations of yellows, purples, and greens, colors determined to preserve their elegance while they inexorably faded into time's dreariness. Now here I was, eating a banana while begging an old man for a plumber.

The Samaras, wealthy enough to be despised by those who were richer and poorer, built their homes along a dirt street curved like a bow beneath the Boulevard. There was nothing adventurous or unorthodox, save perhaps the designs for the cemento. Their favored architecture was meant to convey permanence and the status of the householder, and only fine materials would do.

Marjayoun's stonemasons came in the 1860s from Dhour al-Shweir — a town known for the craft — and from Khanshara and Btighrin. With them came the ideas, styles, and forms of Mount Lebanon, with its array of monasteries and palaces of feudal lords. The aesthetic of the masons was never subtle; they embraced volume, rough-hewn blocks of browns, yellows, and grays. Isber's house, though still dwarfed by the grandest villas in Marjayoun, would be a structure to be reckoned with.

According to his own accounts of construction, Isber loved to watch the masons work. As was his habit, he spoke rarely, but undoubtedly watched carefully to make certain that the work continued at a steady pace. Those who labored around him knew where, when, and how to chisel in order to create the smooth face of the home's façade; they knew where to set the stone to coax out the ancient genius of the arch. Each block was a different size and color. Each had its own story. Fragments and shards of the cut stone littered the hard brown clay of the land.

No house in Marjayoun could lay claim to grandeur without the triple arcade, woven together with a delicate wooden tracery, brought in by builders from the mountains. A debate over what had inspired the arches' design had long continued. Some suggested that the style came from the architecture of Venice, an influence born of a trade stretching back centuries, to when Venice sent salt, wood, linen, wool, velvet, Baltic amber, and Italian coral across the Mediterranean to Egypt, Anatolia, the Levant, and Persia. The East shipped silk, spices, carpets, ceramics, pearls, and precious metals. But the arches were as much a local innovation, as cultures made influences their character.

The grand room behind the arches, the most formal quarter of the house, meant for dining and to receive guests, would be floored in marble imported by ship from Italy, each tile two feet by two feet, white and bordered in black. The work was delicate. In the center of the room was an almost imperceptible design of four pieces of black marble. Overhead was cedar-colored wood, quite expensive in a country that had lost its forests in antiquity. In the other rooms, those meant for family, Isber put the cemento, a mainstay in Europe for a half century but only then coming into fashion in Lebanon. The tiles' geometric patterns, richly rendered, were inlaid into the body of the tile. Produced by skilled workers using a cast-iron mold and hydraulic press, the cemento came in colors brought to life with ground marble dust, fine white Portland cement, and natural earth pigment. The patterns were typically imported from Europe.

Because Isber had traditional taste, many of the tiles drew on a simple design of black stripes, arranged diagonally or along the border. Yet he allowed himself an indulgence. Another design featured in the house was more ornate and included rectangles arranged side by side which resulted in a three-dimensional effect. The use of two patterns was unusual, but the combination converged to create an effect almost dizzying. The design seemed alive, with the pattern seeming to rise from the surface or band together like latticework. A small flower adorned each corner. Overall there was the suggestion of an inverted fleur-de-lis in four colors, the dominant color being deep purple, evoking Tyrian dye and antique royalty.

Miana Maria Ruth Farha was born on Tuesday, August 21. Hikmat and Amina had returned from Beirut, where they and the newborn had stayed with Hikmat's mother for a few weeks. As soon as I saw Hikmat, I knew he was different, unnerved by the birth in that lull between wars that often haunts the country. During these tense days, two countries that had coexisted, uneasily, inside Lebanon since the assassination of former prime minister Rafik Hariri, on Valentine's Day in February 2005, were colliding, and the messy aftermath was playing out in Beirut.

Between the two sides was almost no common ground, snarled as they were in suspicion and anger, entrenched in a terrain crisscrossed by ideology, sectarian affiliation, and, most important, contending perspectives. Hezbollah celebrated a culture of resistance to Israel; its foes, who ostensibly controlled the government, promoted accommodation, or at least disengagement. Hezbollah's patrons were in Iran and Syria; the government looked to France and a fickle United States as its allies. When it was all tallied, the country was split down the middle, and no one knew how to bridge the divide. Lebanon was being drawn closer and closer to another crisis, as Israel waited and civil war between Hezbollah and the government loomed in the background, like a television turned low. The prospect of further combat left most Lebanese grim but not surprised.

A long siege of death in progress; bombing prolonged over days or weeks of close battles and losses; fear unbroken: We can't see the scars from these traumas or how far or where the impacts have penetrated. In the comfort of their living rooms, Americans see pictures of disaster but are routed toward new fronts before sympathies develop or questions become too complicated. Television and the craft I practice show us the drama, not the impact, particularly if the results are subtle and occur or become obvious after the cameras and reporters with their notebooks have left. Our tendency is to consider the resolution of the battle or the war or the conflict, not to take in the tragedies that outlast even the most final sort of conclusion. We never find out, or think to ask, whether the village is rebuilt, or what becomes of the dazed woman who, after one strange, endlessly extended moment, is no longer the mother of children.

In Isber's former domain, the ordinary has been, for nearly a century, interrupted by war, occupation, or what they often call in Arabic "the events." These are circumstances that stop time and postpone or conquer living. Traditions die. Everything normal is interrupted. Life is not lived in wartime, but how long does it take for the breaks in existence to be filled? How many generations? This is a nation in recovery from losses that cannot be remembered or articulated, but which are everywhere — in the head, behind the eyes, in the tears and footsteps and words. After life is bent, torn, exploded, there are shattered pieces that do not heal for years, if at all. What is left are scars and something else — shame, I suppose, shame for letting it all continue. Glances at the past where solace in tradition and myth prevailed only brings more shame over what the present is. We have lost the splendors our ancestors created, and we go elsewhere. People are reminded of that every day here, where an older world, still visible on every corner, fails to hide its superior ways.

We must go, Bahija told the children, fearful that the marauders might indeed burn the house to the ground. Raeefa must have sensed from the time she woke up that this was a day far from ordinary. Her mother, who cherished order, had broken her routine of rising early to begin the Turkish coffee, and the vegetable garden went without Bahija's hand. Even the tomatoes in their neat rows, usually treated like living creatures and, by habit, watered five times a day, were ignored.

Isber's house, conceived and built in splendid times, had opened its doors to another place and era. In Marjayoun and other towns stricken by famine and unrest, the tumult after the Ottomans' departure continued into the years that ensued. After the Ottoman garrison fled Marjayoun, villagers from Khiam, Ibl al-Saqi, Blatt, Dibin, and Khirbe had attacked and looted the Serail. In search of wheat, they began to plunder statelier houses. Many who resided in these places, if ambulatory or observant, left their treasures to save their lives.

At the end of his life, if he looked hard enough, Isber Samara might have been able to stand on his balcony and imagine the route taken up the side of the landmark mountain by Bahija and their children, on the day, in the midst of anarchy and increasing violence, they were forced to leave their new home without him. Of course he had been absent. As usual, Isber was in the Houran, working. Since the completion of the house in Marjayoun, he had grown bored with the town and its provincial snobbery. He had no notion of the danger to his family until the threat had dissipated.

The Samaras had just moved into their house, but it was not the survival of the place or its contents that concerned Bahija. She knew that she would have to save her children, and she did, not waiting for help

*from her husband or anyone else. Without delay, Bahija gathered up
what they could carry and, before departing, assigned each child a task.
Raeefa carried the clothes and was warned not to let them drag. Nabiha,
Isber's oldest daughter, carried Najib, his youngest son, on her back, a
memory she would take with her through all her days. Bahija carried her
daughter Hoda, the youngest. They fled to Rashaya al-Fukhar, ascending the Arqoub region of Mount Hermon, where they stayed until Bahija
deemed it safe.*

Yes, Hikmat was different. He moved a step slower, more tentatively. I
could see the stress, the questions darting across his tired red face, the
anxieties over what his daughter's world would be like, about what this
place would make of her, if it and she survived. He seemed burdened,
and he complained of pains in his leg, hip, and neck.

I asked him how he felt.

"A father," he answered.

Amina told me he was afraid to hold the baby for fear of breaking
her. I remembered that feeling with my little girl. Hikmat shrugged his
shoulders and demonstrated how small he imagined his baby to be.
"Like this," he said, holding out his hand.

We looked at pictures of the baby, whose eyes recalled Hikmat's, and
I remarked at her beauty.

"The monkey in the eyes of his mother is a gazelle," Hikmat said.
He chuckled. "She's our daughter. Whatever anyone says, we're going
to say she's pretty."

Conversation drifted from the lunch table — over plates of a len-
til dish known as *mujadarra*, hummus, pickles, olives, bread, French
fries, salad, eggplant, and the requisite scotch — to the couch. Hikmat
opened the shutters and let a soft light flood the room. Hikmat and
Amina lit cigarettes, and Fahima, a friend of Hikmat's family, brought
a tray of bitter Arabic coffee. The light caught the trail of wafting cigarette smoke. I had quit smoking, at least momentarily.

"I don't want my baby going through a war," Amina told me with a
slight Kentucky twang. Hikmat shook his head. Resignation and con-

descension mingled in his voice, and he had a *zaim*'s way with words, spoken with the authority that declares you more worldly and weary than the next person. "All of us were born in war," he said.

Hikmat's politics were mercurial. He often gave voice to a visceral but inchoate fear of the Christians' fate in southern Lebanon, where they were a distinct minority. "Sandbags for the West," he called them on more than one occasion. "It's bad," he said. "I am worried about Lebanon, Anthony. I'm really worried."

Almost every day, the idea of conflict, of more strife, of the Israelis' intentions, of Lebanon's descent into *fawda*, chaos and anarchy, entered his conversation, as the sonic booms of Israeli jets, illegally flying over the Lebanese border, reverberated. There were threats, warnings, indignation, veiled hints of collaboration and betrayal, all broadcast hour after hour on wildly partisan television channels, as more assassinations and bombings confused Beirut, bringing a kind of social paralysis that precluded any notion of next week, next month, or next year. Leaders seemed to revel in crisis, their casual pronouncements frightening and provoking. There was a sense that Lebanon was condemned. But babies like Hikmat's still arrived, unknowing.

"*Mufajaat*," Hikmat said, an allusion to the surprises that Hezbollah's leader, Hassan Nasrallah, had promised his foes, Israel in particular. "What *mufajaat?*"

Amina spoke bluntly, emotionally, inviting someone to argue with her.

"If there's war," she said, "we have to leave."

She turned to another guest, an army officer and friend of Hikmat's named Elie Deek. "Tell me there's going to be no war," she said to him, pleading.

With his index finger he drew a question mark on his thigh.

In the town square, as Isber surveyed his family's future prospects, men debated the merits of America, informed about the country by little more than the letters that had made their way back to Marjayoun from those who had settled there — children and relatives. The letters arrived weeks, months, sometimes years after they were written by the scores of family members who left every year. The envelopes came with money, occasionally no more than a few pennies, and with praise for a miraculous place where streets were lit, buggies moved on iron tracks, and roads unfurled without end. Streets in New York were lined with buildings that climbed higher than anyone could count. The city itself seemed to have more people than all of Lebanon. Few letters ever conceded a moment of failure. Rarely mentioned was the ridicule and abuse some suffered, derided as Greeks, Turks, or Jews. Klansmen down south burned crosses, and there were many complicated manifestations of racism and populist hatred. (Oddly, no one ever remembered an insult in those days that actually matched their ethnicity. That would await a generation or two.)

Few in Marjayoun would ever know about the lawlessness of the Texas oil fields or the fortunes lost in the coal mines of Oklahoma's Choctaw Nation. More often, readers of letters sent home were given a sense of another place, where money came from hard work and brought a life of ease never endangered by distant wars, village vendettas, or the whims of brigands like those who preyed on Marjayoun again and again. The America represented by its new citizens was, like the identity that the nation had carved out for itself, part myth, obviously an ideal.

Isber heard the letters read aloud, listened as long as anyone had a story to tell of any of the places where his relatives and neighbors had ventured. He rarely commented. Spoken words were best when rarely uttered, and Isber did not break the tradition. To say too much was unwise, a belief sharpened by his years as a trader in the Houran. He saw the success of those few emigrants who did return, in Western-style hats with steamer trunks packed with tailored suits and gifts for nearly every relative — silk handkerchiefs, embroidered tablecloths, jewelry. Isber heard from his nephew Aref, the oldest son of his brother Faris, who went to New York with his sister Offa. He heard, too, from his oldest sister, Khalaya, who had traveled with her hard-drinking husband, Faris Tannous, winding up in Oklahoma, across the Red River from Texas. No one had died there. No one had failed. No one seemed disappointed by the oppor-

tunities they found. No one had to flee their new homes. A few returned,
only to complain about what they had left behind.

A little while after I saw Hikmat and Amina, Karim and I met for
lunch, a way for me to thank him for his recent gifts, which included
mosquito repellent and a bar of apricot soap in a pink container. "I
don't like people that easily, but we seem to be getting along," he said
with his confident flair, his sunglasses set on his forehead, as we chose
a table at a restaurant on the Boulevard. "This is a gift from God."

As we sat there, he kept introducing people to me, usually with a
reference to their roots in our family. As in: "This man's grandmother
is a Shadid." Or: "This man's mother is a Shadid. You would be surprised how many people are Shadids."

No conversations in Marjayoun were more common, more authoritatively deliberated, or more steeped in encyclopedic knowledge than
those about genealogy. By that I mean the myriad intermarriages here
that connected everyone to everyone else. While no one insulted another's family, at least in their presence, everyone was keenly aware of
their own family's saintly attributes and everyone else's sinister flaws.

Tracking one's surname was a constant activity: Hikmat belonged to
Bayt Farha, Isber to Bayt Samara, I to Bayt Shadid. The names themselves were clues to the stories of origins, the points of embarkation
to their emigration. The houses whose names ended in vowels — the
House of Samara, the House of Farha, the House of Gholmia — traced
their origins to the Houran in Syria, where Isber and his brothers
worked, even though the Houran was only a stop in a centuries-long
emigration that, if we are to believe the stories that survived, began in
Yemen, then Jordan. These people were known as Hawarna, the plural
of Hourani (itself a last name), essentially meaning someone from the
Houran. For the most part, the families whose names ended in con-
sonants — the House of Shadid, the House of Tayyar — are known as
baladiya, or local, having been in Marjayoun when the Hawarna began
arriving four hundred years ago, in 1613. They, too, were emigrants,
though their divergent arrivals in Marjayoun lacked the singular, epic
narrative of the Hawarna trek.

Each family seemed to have its story. The Nayfas were named after a particularly remarkable grandmother. So were the Farhas, from the Arabic for joy, and the Shatiras, meaning intelligence. The Gholmias, an elder in the family maintained, were named for the revenge they took over the killing of their young man, a *ghoulam*. Ghoulam became Ghalalimah, the plural of the word, then Gholmia. The Deebas, Bayouds, Razzouks, Zghayers, and Salloums are part of the Khereiwish, named for an ancestor, Kharyoush, who had a guesthouse in the Houran, a breadbasket even in Roman times, when it was known as the province of Auranitis.

Another group of families are known as Labaniya, a name that comes from *laban*, Arabic for yogurt, with which they traded with Bedouins in the steppe. One family took the nickname of a famously beautiful grandmother, Zeina. The Tayyars, from the Arabic word for bird, earned their name because they were said to have left their village suddenly, without warning, and without telling anyone.

I never learned the reason for the name Samara, which means tanned in Arabic. Shadid, easier to explore, can mean strong, powerful, severe, hard, violent, intensive, acute, keen, serious, drastic, or great; we can only guess what might have prompted the appellation.

Don't mess with a Shadid, Hikmat warned me, whispering in strict confidence. There were, of course, variations. "Don't tell me the Shadids aren't smart," Shibil had insisted, formulating a notion of the family's mad genius. "The *akhwat* doesn't come from nothing. It comes from thinking too much. That's what genius is. An *akhwat* is a smart person. He doesn't have no brains. He has too many! He thinks too much!"

All these age-old reputations and disputes still colored the town, particularly the divide between Hawarna and *baladiya*. Until recently, intermarriage was not encouraged. Some quarters continue to belong to one clan or another. Remarkably, hints of the older dialects persist. No one took those enduring differences more seriously than Karim. They were a paradigm to him, a way to view the world, and everyone's character could be explained by delving into his or her origin, Hawarna or *baladiya*. At lunch, he was especially chagrined at Shibil's and Hikmat's talk of Shadid craziness, though in fairness they weren't the only ones to tell me the stories. His voice rising, Karim answered them with a vehemence that swirled like an Oklahoma tornado.

"They claim a Shadid is crazy? You know the Farhas are almost all crazy. And the Jabaras. Oh! Almost all of them. The Ablas. They have either twisted legs," he said, jumping out of his chair to demonstrate a limp, "or twisted minds."

We went through the other families, Karim dispensing opinions. By this point, I was enjoying his utter assurance. He was like a soloist, flamboyant, dramatic, and invigorated by his renditions. In a town that had not bestowed on him the respect he deserved, he was finally having his vengeance, as our food, sitting in the dishes served before us, got cold.

· 6 ·

EARLY HARVEST

I HAD LEFT THE olives I had harvested at Shibil's house, and when I returned, I found them covered in a layer of mold. Shibil said it was salt. It wasn't.

Then he claimed it was the product of salt. It wasn't.

Then he said, "Don't worry," the same thing happened to his olives the year before. *So why was he making himself out to be such an expert on olives?* It wasn't reassuring, but more advice followed. Leave them in salt longer, he told me in the tone of a teacher more weary than exasperated, put less water in the pot, don't wash the jars again. At his most maddening, I asked him how long I should wait to eat them.

"Three weeks, maybe a month," he said with glazed certainty. "If you had crushed them, ten days, but you sliced them." He lit another joint. "You have to wait two months at least," he said, unwittingly changing his advice. "At least. At least two months." I should wait till Christmas? I asked him. "At least," he answered, extending the deadline further.

"Sliced, not crushed," he said, sounding wiser and more authoritative, like Sean Connery, in tuxedo, ordering a drink before baccarat.

"You see, sliced, not crushed," he said again, before falling back in his chair and drifting away.

But I am starting at the end of the story.

Olive trees are ubiquitous in Marjayoun, with forests scattered throughout the plain near Deir Mimas, on the way to the Litani River. Glowing in the pink light of dusk, the trees are as lonely as they are rugged, centuries sometimes interred in their gnarled trunks. "This tree is a blessed tree," Shibil told me once as we sat under the two aristocratic olive trees that stood at the entrance of Isber's house, their leaves carpeting the ground in a light brown, like a fine silk rug thrown over the dirt. He paused and shrugged his shoulders, at a loss for words, as if he was asked what faith means. It seems appropriate that these eccentric-looking trees produce the odd, delicious product called the olive. Olives may be the subtlest of fruits, though I had never really considered them until I arrived in Marjayoun, where they are requisite at every meal and often inspire spirited conversations over their origins and attributes (sometimes fantastical). The olive is *malak al-sufra,* after all, the king of the table, and its disposition and appearance must be kept under constant scrutiny as harvest approaches.

Timing, I learned, is everything.

Bayn tishrin wa tishrin, fee sayf tani, it is said here. Between October and November, there is a brief second summer, it means, when autumn sets in and Marjayoun, now emptied of expatriates who had come back to their homes here to flee Beirut's swelter, begins to slumber. The streets are uncrowded, and the town slows down, though the difference is not radical. Gradually the light becomes softer, more hesitant, as the olive harvest — running from the end of October through November — begins along a stretch of land that once ran as far as the roads did.

"Wait for the first rain to pick, even the second," I was told time and again when I spoke of harvesting olives from the two trees at Isber's house. Shibil and others repeatedly hammered home the point, but I had no patience; I was, after all, American. And there was something admittedly romantic about a harvest that, from afar at least, seemed a soothing ritual, full of renewal and healing. I had sat under these same trees at the end of war, and now I could pick their fruit in peace (a relative term here, admittedly). Olives were already littering the ground by October, and each one that fell felt like a loss to me. Impulsively, I began scooping up the best-preserved fruit from the dirt each day. I put a few in my pocket every so often, then collected them in a black plastic bag.

There is something intrinsically aggressive in a harvest. The tree is pillaged, the crop destroyed, the vines stripped. A tarp is set on the ground, and the branches of the trees are beaten with a stick until the trees relent and surrender their fruit. Perhaps I am too sentimental, but I couldn't do that to the two century-old trees that my grandmother once gazed upon, so I handpicked the olives instead, one by one. I purchased a ladder and a light blue tarp, brought a lunch of chicken, grilled tomatoes, cole slaw, and pickles. Then it was time to begin. Early, yes, but who is counting? (Given my neighbors' inquisitiveness, I am certain there were many, actually.) Engrossed in the colors, I tangled myself in branches. The purple hues of the more mature brown olives melded with the tree's bark, their shadows enveloping others. The green of the younger fruit fluttered against the silver tint of the leaves. At every turn, I stretched, trying to grab the olive at the end of a branch bowing gracefully over the street.

The harvest soon took on a meditative quality. Afternoon was the most peaceful part of the day in Marjayoun: Families napped after their traditionally substantial lunches; I hardly heard a sound around me. There were no birds or bugs, no chatter — only the silence of a small place and the wind whispering. I twisted the ladder in every direction, ascending higher into the first tree, dropping olive after olive in the plastic bag tied to my waist. The ladder teetered as I reached for the fruit. The bag eventually bulged. Within a few hours, just a few last olives remained. They were reachable only if I climbed the tree. I did, clambering up branches that were deceptively strong. I leaned and stretched, holding my breath and squinting, then pulled branches toward me, before surrendering and calling a truce of sorts. I told myself again, not everything would be mine. The rest of the olives would wither on the tree as they had for so many years before, dusty and bare, dangling from branches I couldn't reach.

"Those are my aunt's trees!" George Abla, well into his nineties, shouted at me as I relented. He was my great-grandmother's nephew. Wearing a frayed suit jacket and a woven blue hat with what resembled a cotton ball on top, he had stopped to admire the trees from the street, leaning on his wooden cane. I smiled at him.

By dusk I was finished, and I sat under the trees. My forearms, shoulders, and back were scratched by the coy branches near the tree's crown. Dust clung to my beard. The village remained quiet, and the house felt empty and haunted. The diminishing light angled from the west, casting a shadow that unveiled the house's desolation and abandonment, spilling forth everywhere. The reconstruction already felt overwhelming, a feeling exacerbated by Abu Jean's tendency to turn every minor job or adjustment into an adventure or calamity, along with a litany of promises by workers who failed to show up. The house was still a remnant of another time, an artifact in a way. But sitting here, after an impetuously early harvest, I felt a simple connection, even if it was merely picking olive trees that belonged to the past.

"I have 100 kilograms of olives," I messaged Shibil. It was a joke, of course, but, apparently alarmed, he soon pulled up at the house in his Mercedes, parked in his usual erratic fashion, and repeated what everyone else had already said: Only after two rains, at minimum, should I pick the olives. He said it over and over, shaking his head.

I have an obsession with Arabic that comes, I suppose, from learning it as a second language. Words are imbued with both elegance and logic, chiseled by their sense of having moved through time, history, and generations. A beauty ensues, as words slowly unfurl their mysteries, shifting their meanings ever so gradually. A richer language than English, Arabic boasts a vocabulary that can convey any description in a single word. Arabic also has a far greater facility to communicate sarcasm, and it can be employed precisely, or with pitch-perfect irony. Overqualification, pronounced emphatically, deflates an ego in Cairo, as in addressing someone with an undeserved Ottoman honorific — *basha, bey,* or, my favorite, *bashmohandis,* the Pasha Engineer. I called Shibil *ustaz,* a word that can range from "mister" and "lawyer" to "teacher" and "professor." He heard the sarcasm and threw it back at me. I became sir, lord, and beloved, all at once. "*Ya ustaz, ya khwaja, ya habibi,* if you had waited one rain, one time, they would have looked like this," he said, holding out his index finger, marked off by his thumb, the same way he measured hashish. "If you had just waited."

Shibil was a perpetual skeptic, and he delivered the same doubts about the harvest of my two olive trees as he did about the rebuilding of the house they adorned. "What, are you going to hire twenty Syrian workers to harvest your olives?" he had asked as the season approached. The absurd followed the absurd: "What, do you have a million trees? Do you have twenty thousand acres?" As he continued, he seemed to grow more cynical. "Did you call the Syrian president and ask him to send ten thousand Syrian workers?" He laughed boisterously. "Each one will pick one olive!"

His disapproval was probably less ridicule and more an attempt to assert his expertise. In a town where he had little to buck him up, here was something he knew — how to cure olives (and pickle a vast assortment of vegetables, which went well with scotch). He lectured, for what seemed like hours, on the proper cucumber for pickling (small but not too small), the optimum mix of spices (a great many, according to Shibil), and the ratio of salt to water to make brine (enough salt so that an egg will float in the water). My unwillingness to treat him as an *ustaz,* even in curing olives, was an insult, and his ridicule had the same defensive pride I had detected in him during almost every encounter in Marjayoun. He was an expert.

That night, I brought the harvest to Shibil and, not surprisingly, the simple process became far more complicated. Indeed, the dynamics seemed intangible, arcane, and decipherable only by him. In Shibil's disquisition on the art of curing olives, I was left with a mix of nuclear engineering and Sufi mysticism. The sheer mystery of the procedure prompted more questions, which led him to offer more esoteric answers. Despite his moments of lunacy, Shibil was endearing, especially given his stoned demeanor. But whatever I did, he corrected me, again and again, in withering fashion.

- *Washing:* I had mistakenly left in the rotted olives I had picked off the ground, imperiling the entire batch.
- *Curing:* I had ignored the fact that there were two processes, one for green, one for black, both of which were harvested from my trees.
- *Mixing:* I had yet to comprehend the subtleties of *tosht al-bayda*, by which you measure the amount of rock salt to put in the water (again, enough so that an egg will float).
- *And other details:* In instructions conflicting, colliding, and contradicting, I was told (corrected) about everything from where I should set the olives to dry to how long they should soak in water before curing. This dialogue went on for days.

Unexpectedly, we finished the olives after finding the orange, lemon, and bay leaves we would put in the jars. We cut slices of lemon for some, red pepper for others. We would add the olive oil along the rim later. "Why wait for the oil?" Shibil asked me pedagogically, as we worked a week later in my tiny kitchen, standing on a weathered brown Persian carpet I had bought at an auction in Baghdad years before. "To let the salt and water penetrate the olives."

With that, we were done at last. My kitchen countertop was crowded with fourteen pickle jars full of olives. They didn't seem too small an amount despite my early harvest. I kept glancing at them, holding them up to the light, seeing if I could notice any change after a few minutes, a few hours, a few days. Through the looking glass, the slice of lemon drifted to the bottom. The salt crystals, reflecting the light, sat on the fruit, as if hoping their presence would be ignored. I marveled when I noticed the perfect olive, and there were only a few in the jars — an oval with none of the ravines, gullies, and valleys of the more weathered ones. I wondered what it would taste like, as my tongue rolled around the pit.

"The first batch from Bayt Shadid in generations," I said proudly.

"With more than a little help," Shibil answered. "*Maakul al-hanna*," he said. "Eaten gracefully." Then, skipping across languages, he added, "Bon appétit."

• • •

I was lonely in Marjayoun, yet despite all the discussion of Lebanon's internal strife, the longer I stayed there, the farther away war seemed. I often pictured my daughter Laila walking past the stone wall, up the buckling driveway, and toward the antique front door I was determined to save. I thought of the day I would bring her here, to a house she could call hers. Fog began to roll in these nights, and as the haze erased any definite sense of time, I thought of Isber, too, a determined man nodding his farewells at the houses of his youth and the mountains above as he made his way through a town once as prosperous as he was.

Isber Samara was becoming an acquaintance in a way. We shared something, I suppose, perhaps more than ambition, and I marveled at the fact that he built this house that was an artifact of a Middle East gone but for its power of inspiration. I felt more at home in his town, where I imagined Bahija in the kitchen, spreading sugar and butter on fresh bread for her children. During those autumn weeks, I often thought of a line that Shibil had told me: "Be with folks for forty days, and either you'll be a part of them or you'll leave." More than forty days had passed.

I was learning the basics of getting by, the rules and traditions of living among the townspeople. A rule: I should never tell anyone how much I paid for anything. If I did, I should discount it fifty percent. Nevertheless, whatever the sum, I was greeted with the inevitable *harram,* "what a pity," always spoken with the greatest sympathy at my naiveté. The manipulation of guilt, in matters of family and friends, was an art form. Or so I learned. In Marjayoun, I was subjected to an aggressive variety; no matter how distant the relative, I was always being pummeled. *Why haven't you visited this month?* the question went. If I had, it became: *Why haven't you visited this week?* If I had gone that week, it became: *Why didn't you come earlier today?* Or: *Why are you leaving so soon?*

Karim had his own variation. Anything short of making his house the first stop on any return from Beirut or elsewhere was simply unacceptable, particularly if any time had passed without contact by telephone. "You haven't called," he would tell me reflexively. "Are you cross with me?" The question — *why did you beat me as a child?* — could not have been uttered with more poignancy or desperation. It turned out that *he* had called, four times — twice from his house in Beirut, twice from a kiosk on the corniche along the Mediterranean, where he did his fast walking. When I saw Shibil at my rented apartment, after a couple of days of no contact, I tapped my forty days of knowledge, courtesy of Karim and others. I was determined to turn the tables.

"Where have you been, *ya ustazi al-aziz?*" I asked indignantly, with a look of disbelief. He fumbled for an answer. It was my first Jedeidani guilt trip, entirely satisfying.

Abu Jean and I had very different notions of the job and the work it entailed. I saw him as the foreman, who would commandeer the project and wrestle it, through force of will, to fruition. Abu Jean thought that by taking on the project he was doing a favor for Hikmat. I realized that if we ever finished the house, Abu Jean and I would have to become some other sort of team. We got on most of the time. I wrote phone numbers down for him in the ancient notebook he kept in his back pocket. Then Abu Jean would call, asking me to read the digits out loud and slowly. Every few days, he would pile in my car and we would rumble over the potholed roads to Qlayaa and Kfar Killa in search of a carpenter, to Bwaida for a cement maker, to Khiam for stone, tiles, and other supplies. The blacksmith, electrician, and tiler were in Marjayoun. So was the plumber, hanging out at his cousin's shop, smoking cigarettes, as if he was still in high school.

Abu Jean had a car, a 1960 gray Peugeot, but he never drove it. He was either afraid to or employing a bit of *shatara*, or cunning: Why pay for gas when Anthony will drive?

On bad days, Abu Jean and I fought. Mostly, I tried to walk away, cognizant of the insult it was for a younger man to quarrel with someone older. (Hikmat told me that young men on Marjayoun's streets used to put out their cigarettes if an elderly man happened to walk past.) The curse of Bayt Shadid lingered, though.

When I lost my temper, and I did every couple of weeks, Abu Jean would storm off, usually coming back the next day, indignant and describing to me the aggravation he had endured the night before, as he

sat at home replaying the events in his head. I would apologize. Bayt
Shadid, I would say, by way of explanation. He would nod knowingly.
It was a language that he understood. After one apology, he confided
that his family, Bayt Abu Kassem, had a reputation for craziness as
well. I didn't dispute the possibility. Bayt Shadid and Bayt Abu Kassem,
both mad, would manage an impossible project together. I hoped.

"I trust Abu Jean one hundred and ten percent," Hikmat told me.

To Hikmat, Abu Jean was a paragon of honesty, and his loyalty ran
far deeper to him than to me.

"From the days of my father, there's a bond with his family," he said.
"His kind of people are no more. They don't know evil yet. They have
virgin spirits. You think there are a lot of people like Abu Jean in Lebanon? Everyone in Lebanon is a Mafioso man. Unfortunately we have
in Lebanon only Abu Jean. Or maybe two or three Abu Jeans."

Hikmat was an authority on construction, or at least saw himself as
such, having rebuilt his house years before.

"Everyone has an opinion," said Hikmat, who still bore the scars
of his own reconstruction, speaking not only of workmen but also of
neighbors, acquaintances, distant cousins, and passersby. "It will kill
you. You'll be like an idiot, sitting between them, and they'll be throwing opinions at you." From the time the workers started pouring the
cement, people came to watch. "Do it this way, do it that way," he recalled.

Finally, the foreman shouted at all the hangers-on, "You sons of
bitches! Out! Out! And shut your mouths, too."

A day after lamenting the way people yelled in his ears, judging him
an idiot, Hikmat joined me at the house, where he proceeded to treat
me like an idiot. With Abu Jean and Amina, we ambled past the old
garage gate, rusted, mangled, and ripped from the wall, then gingerly
made our way into the house, past menacing piles of dirt and masonry.
A pool of dark gray cement gathered on one floor, with a shovel sticking out of it at a forty-five-degree angle. Near it were three bags of cement — one untouched, the other two torn open — and a white plastic
bucket filled with water. The Baghdadi — an ancient amalgam of mud

and straw that was once on the walls — was piled elsewhere. It was mixed with wood, plaster, pieces of roof shingles, even an old chair. Rickety but resilient, the chair seemed as if it might last forever.

Hikmat turned to me.

"I thought it was going to be worse," he declared, looking serious.

I laughed, cocking my eyebrow.

Amina sensed Hikmat's ploy, a brazen attempt to instill confidence in me through sheer denial. "I thought it was going to be better," she said, with a little more concern.

Hikmat ignored her.

"This is easy," he said assuredly. "If you know what you're doing, it's easy."

He looked around at the stone walls, moving toward the dining room. "Tell Fouad and his brother, this is easy." He paused. "Tell Fouad, it's a very easy job."

Amina kept bringing him back to the reality that he was determined to subvert.

"You've got your work cut out for you," she told me.

Hikmat roamed the house, all the time haughtily seeming to consider himself the one in charge. "Don't get rid of the metal," he said, pointing to the grills that spanned the windows, then moving to the dining room. "Put the bathroom here," he said. Despite his bossiness, I believed he was sincerely attempting to be of help. "Don't close the window in the kitchen," he ordered, stretching his head into another room, and ruling, "You don't need any more light. That's plenty." Later he reappeared, still in take-charge mode, his certainty inspiring to my flagging spirits. "Put parquet in the Cave," he advised. "Fuck everyone else and their opinions."

As we walked to his car, Hikmat again put the blame on Fouad, my cousin by marriage and the engineer. "Fouad, my ass," he said. "Tell Fouad, any delay, it's because of him." With those words, an epiphany came. I now understood the relentless logic underpinning everything that had transpired as Hikmat sauntered through the house, the drama of the exposition. *Given:* Hikmat recommended Abu Jean, a friend of his family, for the job. *Thus:* Abu Jean's failure reflected poorly on Hikmat, indicting his judgment. *Given:* Fouad is from outside the town. As such, Hikmat could care less what Fouad thinks about him. *Thus:* Put the blame for any failure on Fouad, one of the most decent, kind, and generous men I had met in Lebanon.

The thinking was linear, precise, and unforgiving.

Fouad, my ass.

7

DON'T TELL THE NEIGHBORS

A S IT TURNED OUT, Isber's house, by virtue of his genealogy and its geography, was susceptible to prying. My neighbor Wissam was a relative whose home was built by Isber's brother Faris. His mother was my grandmother's first cousin, another tangled tie of blood, but that meant little to Wissam. An unfriendly, bald, pipe-smoking sort, usually gruff and monosyllabic, he had greeted my arrival by making off with purple plums from a tree near the house's entrance. When he had glimpsed me staring in disbelief, he brought them to me, still chilled from his refrigerator. His excuse: He had taken them only to wash them. There was an awkward moment as he waited for thanks. I offered none, and his dislike for me intensified.

These days, he was livid at the dust that the reconstruction sent swirling to his property. The dirt prompted a string of requests. Could I hire a maid to clean his house? Could I clean the red tiles of his roof? And while I was at it, could I repair his crumbling stone wall along with mine? All at my expense, of course.

Massoud Samara owned the land behind me. In Marjayoun, there is always tension about property, any kind. So how could I have been surprised by what occurred?

"Where does my land extend to?" Massoud asked his surveyor one day as I stood with them. He asked it repeatedly, and each time I was filled with more dread, worry, and apprehension. Will he encroach on my land? I wondered. Is he going to lay claim to more of it? Will he try to build a wall over land that is not his?

The questions multiplied. Then Massoud and his surveyor suggested I chop down a sprawling fig tree along the edge of my property. I cringed. I told them that I doubted I could, and I repeated it three times. "This is from my grandmother, from her day," I pleaded. The sentiment was lost on them, but after a few minutes of back-and-forth, they judged it better to delay the inevitable confrontation.

With each day, I saw more of Marjayoun's quirks, the often forgotten, sometimes maddening, occasionally endearing details of everyday life that revolved around my neighbors in Hayy al-Serail, who alternated between effusive warmth and stony, suspicious silence. Always there was a suggestion that I had money to spend. There was a gamesmanship to it all, the jousting of a town that seemed to relish, even celebrate, such contests against a backdrop where war might lurk.

The Ottoman pashas and beys had faded into a landscape never settled since. Anarchy became bloodshed, uncertainty, insurrection, departures. Old animosities returned to curse each other, augmented by imperial manipulation. Gangs exploiting the new lawlessness were joined by bandits, fired by nationalism, who brought a new incarnation to the Arabic word fawda, *or chaos.*

In Marjayoun, Isber and his house were small distractions from the troubles.

A way of life born of the Ottomans, a style of living that was never before especially fragile, had broken into pieces jagged and dangerous. With no one stepping in to replace the Ottomans, or allowed to, other than the Europeans, the Middle East was unraveling, especially in the hinterlands. What had given the region its only sense of identity was gone, and there was nothing to fill the vacuum. Order was breaking down.

Yet Hayy al-Serail was lush and still, with no disruptions to trouble the order of the day. Limbs of old trees reached across streets, and in the mornings, fog drifting up from the river pleasingly obscured anything that the distinguished residents of this tranquil enclave preferred to ignore. What would it take to disturb this place? More than Isber Samara, the newcomer, who would never be more than that for those who lived here, families such as the Farhas and the Barakats, whose forebears seem to have arrived with the ancient olive trees.

Upon its unveiling, the Samara residence, envisioned and reenvisioned through all of Isber's years of waiting, revealed a place that, its owner decreed, with the special pride that accompanies an active imagination, would never be completed. Artisans and craftsmen had done their best work, creating an elegant synthesis of all the signatures of Levantine life. Although for a brief moment the reflections of the candles from Isber's table stilled fears and brought calm, the house would not remain immune to what was taking place outside. It would change with the landscape that surrounded it as Isber became a different man, a family man, a man less focused on his own ambition and glory. His house was first a display of pride, then a place where he made a home for his wife and children. It would in time become a refuge, and finally a memory we carried, whether we ever stepped through its doors.

That night, I saw Shibil, and over a drink I told him about my encounter at the fig tree. I was speaking to Massoud Samara in Arabic, I said, and I thought my words might have been less diplomatic than they would have been in English. I recounted my plea — leave the tree, it hails from my grandmother's time — and asked if I was perhaps too aggressive.

"Aggressive?" he barked. "Aggressive?"

He shook his head.

"Here's aggressive." Shibil sat up in his chair, bowing his back. "Take your hands off the tree," he shouted, "you brother of a whore, before I fuck your sister!" He paused.

"You brother of a whore!" he shouted again.

Shibil flashed me a satisfied look, the same one he had given me when he got the salt just right in the olive jars.

"Sister" is better, he assured me.

"I wouldn't want to bring in the mothers at this stage."

My problems didn't stop with the neighbors. To much of Marjayoun, I was unhinged, at best: The work at the house translated into my being considered certifiable. Some suspected that I was fabulously wealthy. Others deemed me truly, even dangerously insane. A substantial portion seemed convinced that I was a spy, my cover brilliant. According to one rumor, the American embassy in Beirut was paying for me to rebuild my house. It turned out that the more forceful proponents of my clandestine life were Karim's elderly women friends and the Deeba brothers, who owned two grocery stores on the Boulevard.

The unsightly, ogreish, eldest Deeba brother had refused to speak with me, on principle. He would often scurry into the back room when I entered his store. Any word, he probably feared, would find its way into my cables to the embassy. I eventually quit going to his store, settling on the smaller, more modest one owned by the loquacious youngest brother. He was George Deeba, failing merchant and failed politician, who bore a striking resemblance to Humpty Dumpty on a strict diet.

As I entered, he invariably shouted a nickname: "*Ustaz Shibil!*" (remembering me as Shibil's friend) or "*Washington Post!*" (remembering that I worked at a newspaper). It was all a cover, of course. Like his brothers, he was firmly convinced of my role as an agent and determined to discern my mission. He would show off a few words of English, which he last spoke as a schoolboy thirty years ago, now unintelligible, then offer a thought, prompted by nothing. "You've come to study our climate!" he said one day, a gesture either to that evergreen of conversation or to his latest conjecture of my task in Marjayoun.

We chatted about what I was looking for — rock salt, which he had, and bay leaves, which he didn't. He took my number, suggesting he would call me if he found them. He soon played his hand, though. Could I put an announcement about his political ambitions in the newspaper?

That would cost thousands of dollars, I told him.

"Oh!" he shouted. "Not for free? Could you do it for free for me?"

He said he just wanted two lines in the paper, nominating himself for the presidency of the Lebanese republic — the candidates would be chosen in a few weeks. "Why not?" he asked me. "If swindlers and con men can be president, why can't an ordinary guy?

"I have a beautiful program," he declared. "Twenty-five points, and there might be more, but these are ready." They were scrawled on both sides of a memo pad embossed with the logo for Winston cigarettes. "This is called a program," he said, shuffling through the four pages, quickly scanning the points.

I asked him the most important of them.

"The most important point?" He kept looking through the pages, turning them over and over a little obsessively. He wrinkled his brow, growing confused.

"They're all important, all twenty-five are important." He kept reading, stopping on one point. "Well, this isn't that important," he admitted. "And this one," he went on, "really matters only for Lebanon." His finger went down the list. "This is important internationally." He paused. "This one probably just for the region."

He shrugged his shoulders, tucking the papers back under the counter.

As I left, I said, "Thank you, *ustaz* George."

"George W!" he shouted back at me.

As time passed, my collaboration with Abu Jean bore fruit, though there was no specific breakthrough. Everything happened by default, and eventually we stumbled on the *maalimeen,* the contractors, who could work on the house. Before long, they became a community, the fractious, divided, and dysfunctional sort that finds itself susceptible to civil war.

The mainstay was Toamallah al-Qadri, better known as Toama. With his wife, Thanaya, he lived next door in my cousin Wissam's house, paying $100 a month for a modest three rooms that faced the *warshe,* or workshop, which had by now become the nickname of Isber's house.

Fleshy, with an ambling gait, Toama would do anything. But everything — including the cost of gas for his decrepit red Mercedes when he went to check for replacement stones — came at a price. (The gas for the errand cost about $16.) He was a hustler, good and bad, and Abu Jean kept a wary eye on him.

Emad Deeba, the electrician, a proud man rare to smile, was the next to join the crew, followed by Fouad Abla, the plumber made weary by war, and, most memorably, Kamil Haddad. No one called this man of bounteous gray hair and pale blue eyes, limpid and almost transparent, by his first name. He was simply Abu Salim, one of the last true stonemasons left, a handsome man whose face was as chiseled as the stone he worked with for six decades.

While Emad the electrician lamented the amount of work — "it's a big job," he told me gloomily — and Fouad the plumber warned me that other projects would take precedence, Abu Salim strolled around the house with the authority that came with swinging a sledgehammer and still one-arming a stone equal to his weight. Did I mention he was seventy-six, as old as Abu Jean? For $1,100, he would build a stone wall over the garage entrance, turning it into a room, enclose a window with stone, shrink three others, enlarge a fourth into a door, and repair the Cave. I agreed; I dared not bargain. The next morning, he was at work, arriving at 6 A.M., the sheen of dew on the cars yet to dry.

"Be careful, be careful," Abu Salim told his apprentice as they pulled on a window frame. "Treat it like a bride." Soon Abu Jean ambled into the *warshe.* "Whatever you need, Abu Salim, I'm at your service." Abu Salim answered him, with the grace that is so entrenched in Arabic, "I want your peace and safety and nothing else."

In time, all three men were working on the wall, with a pickax, a crowbar, and a sledgehammer, removing the stone below the window to make way for the door.

Abu Salim began shouting as he bludgeoned the wall. "This is the foundation of building!" he cried out gleefully. "The stone won't go anywhere!" He was jubilant at his lack of success. Building, after all, is an art, and he was appreciating a masterpiece. "This is the old style of building! This is the way homes were built!" he said.

"Building is the most important art of life on earth," Abu Salim told me solemnly.

"This type of building is going to soon become extinct, and it's a shame," he went on. "It's a shame because it's from the earth, it's the earth's yeast."

Abu Salim told me the names of the tools he used. "Each one has a specific function," he said, a little didactically. "You can't use them interchangeably."

There was the *tartabeek* (a kind of pickax), *mahadda* (a big sledgehammer), *makhal* (crowbar), *fas* (ax), *mankoush* (another pickax), *shauuf* (a kind of mallet or chisel), and a *shaqoul* — a graceful way to measure whether a stone is level, a sort of plumb bob. His was a green rope with wood on one side, a brass weight on the other.

He looked at the plumb bob, again with that sense of appreciation. "This is hundreds of years old. This was inherited from my father." I asked him if it was his most important tool. He laughed, then looked at me as though I was either a fool or too young to comprehend such things. "The hands are the most important tool!" he shouted.

"Look, *ammi,* you want to know the truth?" Abu Salim asked me a little later, as we sat outside the house under a warming morning sun. "Back then was purer than today. The atmosphere, the aura of people. There wasn't this hatred, this revenge that we have today. People lived honorably, simply. There was no electricity. They lived on candlelight, by the oil lamp, and that's it. They'd have each other over for dinner, and enjoy their evenings together. They'd offer one another figs, roasted chickpeas, and raisins. You'd wake up at seven A.M. to go to work, and you wouldn't come home until nighttime. There were no worries, no wars, no fear for your children. It was simply another time."

Abu Salim worked with the rock's very nature, understanding that he could never use the brute force of a machine to force it to submit to his will. It is stone, after all, earth's yeast. Instead of doing battle with it, he would coax it, persuade it, nudge it. The more experience, the more perfect his knowledge of the stone, the more innate his sense of what would follow — where the stone would splinter, how it would splinter, and when it would splinter. As I watched him, it reminded me of something I once read about the most practiced surgeons — how they could so quickly predict the consequences of their hands' every move. It was the same for Abu Salim.

He went to work removing another stone, a foot at a time, probably fifty pounds or more. Both he and Abu Jean appreciated the mortar and how tough it was.

"See how this is?" Abu Jean said. "This is the old cement."

"*Shou lakan*," Abu Salim answered. Of course.

They were two men near the end of both life and career, interchangeable here. They knew their work, the angles and foundations.

"This is construction!" Abu Salim shouted again. "What you give, it will give. What you take, it will take." The old man seemed triumphant, even euphoric. "The strength of the old buildings was neither iron nor cement. It was stone. And the strength of the men back then came from honey and the fat of the goat."

The work went on through the morning. He rained a hundred more blows on the stone, from the level of his waist and crashing down from above his head. He was a marvel to watch, one hammer blow after the other, without break, machine-like. As he toiled, he recited poetry: "The artist never rests. The stars of the sky sing to him. My son, don't blame the old one. In the past, there was nothing like him."

"Look," he said later. He shook his head in forced modesty at his own mastery. "The beauty of work is in its perfection."

One day, after a trip to Beirut for supplies, I had gone straight to Shibil's for a drink. I asked how he was. "*Zift*," he told me, using a word that can mean asphalt or shitty. In Shibil's case, it meant the latter. "Everything is *zift* in Lebanon," he said, "except for the streets."

When I wasn't at Isber's house, I was spending most of my time with Hikmat, Karim, and Shibil, whose relationships were growing more complicated. They had hardly talked of late, but when Hikmat's daughter, Miana, was born, Shibil called him in Beirut.

"I said to Hikmat, 'Congratulations. Say hi to Amina. I'll see you when you come.' *Bas.* That's it. Very friendly." Shibil paused. "We don't have anything between us, but Hikmat doesn't care about anyone. He doesn't ask about anyone. I have no grudge, but why does Hikmat's head have to grow?"

After Shibil served me a scotch, the conversation returned to Hikmat, as it tended to. "I was thinking about visiting him," Shibil said. He had wanted to congratulate Hikmat in person and bring a gift for his daughter, but he'd overslept, his nap stretching from 3 P.M. until 9:30. A shirt that read *In Oklahoma, Nothing Tips Like a Cow* was stretched taut across his belly.

When I told him Hikmat had gone back to Beirut, he let out a string of expletives. "Maybe I'll end up visiting Hikmat's daughter when she's engaged," Shibil said, and I laughed.

Karim, for his part, was never satisfied. Why hadn't I called, he would ask every time we spoke. He had sat alone, going over the reasons. Was it that he had asked me to water his plants? Had he said something wrong to me? Was Hikmat poisoning me with falsehoods?

He gave me yet more presents, in two white plastic bags with pink polkadots: Calvin Klein body moisturizer (Escape for Men), shave gel (Men, Carolina Herrera), shower gel (212, Carolina Herrera), and a black alarm clock, already set to the right time. Stuffed in the other bag were mosquito tablets, a yellow ashtray, a brush to scrub the sink, and a small strainer to steep tea leaves. It was either a housewarming gift or an attempt at apology, since last time I saw him we had argued about his relentless criticism of Hikmat.

In his eyes, I had lost my cool.

"This is the Shadid in you," he said knowingly. "The temper."

Karim was a sensitive, obsessive soul.

"This fucking town is a mess when the gossip starts," Hikmat told me soon after. "It's not your business to take sides. You're a guest in Marjayoun."

All this, Hikmat added, also applied to politics, which ended countless friendships here. Just a generation or two before, the town had been disposed to hosting movements, ideologies, and caudillos who claimed to represent them. The time was heady, as the Middle East emerged from Ottoman rule and colonial hegemony. The newspapers in Marjayoun favored either the Arab nationalists or liberals. Then there were the Communists, who flourished in the 1930s, and the Syrian Social Nationalist Party, founded in secret, in 1932, by Antoun Saadeh.

In the days of Hikmat's father, the town was beholden to those Syrian nationalists and the Communists, who organized among Marjayounis of modest means. The two factions despised each other. After a few drinks, loyalists of each party would pour out of a bar on the Boulevard and slug it out, with fists rather than guns, well into the 1960s. Little joined the two except for their espousal of a universal standard of citizenship, unusual in Lebanese politics until today.

Some families in Marjayoun were still known for their now faded loyalties to those parties, withered as they were. Bayt Sukkarieh, Bayt Shammas, and Bayt Shambour were known as Communists. One Mujalli Shammas actually painted his house red. (In later years, the party in Marjayoun would split into two branches, Trotskyist and Stalinist.) Bayt Shadid, Bayt Tayyar, Bayt Khouri, and Bayt Musallam were the Syrian nationalists' traditional supporters. (Shibil pasted the party emblem on his phone.) Of course, after the civil war, their numbers dwindled, as did ideological fervor; most of the old party cadres passed away or gave up on politics. More and more, people in Marjayoun and elsewhere identified themselves by religion. They were simply Christian, like my family, and for many Christians here, their leader was a brusque former general named Michel Aoun, given to quixotic endeavors. ("When somebody provokes me, I say, 'Go to hell!'" he once told me.) The identification reinforced their minority status, and I always suspected their specificity probably meant their eventual extinction.

The ideologies in the days of Hikmat's father were about contesting frontiers — Arab nationalism, pan-Syrian nationalism, and communism, all imagining a broader community on the terms they would establish. Artificial and forced, instruments themselves of repression, the borders were their obstacle, having wiped away what was best about the Arab world. They hewed to no certain logic; a glimpse at any map suggests as much. The lines are too straight, too precise to embrace the ambiguities of geography and history. They are frontiers without frontiers, ignorant of trajectories shaped by centuries, even millennia. Marjayoun suffered with the advent of borders, losing its true hinterland in Palestine and Syria and all the more accessible towns there. Those towns of an older antiquity — Haifa, Acre, Jerusalem, Damascus, and Quneitra, a boomtown nestled in the Golan Heights — shared with it a common geography, history, trade, and culture, unfettered by borders, and for generations that land was the place of opportunity for those who chose to remain in Marjayoun. Now they no longer could.

Gone was what had redeemed that long-ago Ottoman era, a Levant of many ethnicities and faiths that managed to intersect before the vagaries of nationalism. Myths had to be imagined to join a certain people to a certain land that was so long shared. Pasts were created, and destinies claimed. The borders reinforced the particulars of states with no ambition save the preservation of a petty despot's power, or a people's chauvinism, or a clan's fear, and cosmopolitan cities gradually but irrevocably became national ones. In the century that followed the fall of the Ottoman Empire, all those states failed; none would quite capture the ambitions or demarcate the environments of the diverse peoples who had lived there so long.

And, of course, there are no more street fights in Marjayoun.

8

ABU JEAN, DOES THIS PLEASE YOU?

T HERE WAS A BOXY, bedraggled red Renault parked across the street from my apartment. Every morning between 7 and 7:30, its engine was revved. It screeched as the gas pedal was pressed. It squealed and howled, in a squall of obsolete engineering, wretched manners, and questionable judgment of the engine's needs. I stared at the ceiling over my bed, eyes growing wider.

This meant I woke up early on these chillier mornings in Marjayoun. November was ending, with autumn holding on with tenacity before surrendering. Old men, smoking and chatting in chairs along curbs each evening, retreated inside. Stores closed earlier, as the steps of the few people still walking the streets slowed. A lumbering, ancient blue Mercedes arrived to hawk with honks the dozen carpets tied to its roof. More piled in back were destined for floors with tile often too cold for even the determined to walk on. Around Hikmat's house, I had faintly smelled the small logs that burned in winter stoves. Smoke unfurled from chimneys, the narrow stovepipes snaking out windows and portals cut through stone. On the horizon, snow tumbled down the crests of Mount Hermon.

By now, the house had started to come to life; I could imagine lives playing out there. I lavished thanks on the men, drawing incredulous stares. To concede that someone's work is good is to risk a possible bargaining opportunity, though I could no longer accuse them of not doing enough. There were five workers at this point. That is, if you count Abu Jean, who, like all the great divas, was frequently unavailable. Overwhelmed by the pressure, harried by the world's expectations and, of course, his own, and bewildered by all the clatter, he had no choice but to withdraw occasionally, disappearing into a cloud of dust.

Abu Jassim and Mohieddin, two workers from Syria, were chiseling old plaster off the inside walls, carting it past the olive trees, and hacking off concrete that covered a stone wall behind the house. Cement was poured to buttress the roof. Still reciting poetry, Abu Salim enlarged the window in the kitchen and made the one in the bathroom smaller. In what was once the garage, a scar on the house and the universe itself, he restored the old walls, all the while considering the play of the shafts of light that had returned to these halls.

Building another wall, Abu Jean hit each cinderblock with a hammer from above, then a lighter tap or two on each side. With a flick of the wrist, timed perfectly, he dropped a scoop of mortar with an accuracy remarkable for his age, then smoothed it with the swift motion of a trowel. He lifted the blocks to where they belonged, hit them first with his fist, then the hammer, then the trowel again, to budge them into place. A yellow string was pulled taut across the wall's façade as it rose. If it bulged or sagged, Abu Jean knew he had lost his angle.

"See the work?" he told me. "That's work."

He shook his head, tired but satisfied.

"What more can I do?" he soon asked, but I had to answer that there was nothing.

"This work, and I'm seventy-four?" he said to me, turning once more to the last block skillfully attacked. A few seconds passed. "No, wait, I'm seventy-six."

He thought for a moment. "I was born in '31," he said, nodding again with a bit of certainty.

"You look fifty," said Nassib Subhiyya, the blacksmith, but Abu Jean didn't hear. So I repeated the praise, louder, leaning toward his ear. "*Taqbourni*," Abu Jean told me: May you bury me. Then he ran fingers, dusted in gray, over my beard, tenderly, his fingers and thumb coming to a point at my chin.

So little of Marjayoun's history was gentle. The place is scarred, as is so much of the land around it. But as autumn approached winter, a community had emerged at the *warshe*. Faces had become familiar; circumstances even I deemed unlikely brought a certain ease that was so foreign in Beirut, prickly as the cast of characters was there. Nearly everyone in the capital ran through a taxi driver's battery of questions to determine a newcomer's religious sect. First name? If it was ambiguous — say, in the case of Nabeeh — then the last name was sought as a clue. Often that was not specific enough, as in the case of al-Hajj. So queries turned to the names of relatives, starting with the father and mother. If they were, for instance, Nabil and Nada, an inquiry into one's hometown would follow. If it was a mixed town, maybe Aley, in the Chouf Mountains beyond Beirut, you might hear the question "How do you see Walid Jumblatt, the leader of the Druze?" or "Are you a partisan of Jumblatt?" There was a relentlessness to it, as a friend once told me, this "categorizing and oversimplifying."

It was present even at the *warshe*. Christians sometimes dismissed designs or color patterns as being too Islamic. Trust, especially when it came to money, bore a sectarian stamp. But these days at least, the categorizing felt softer and more remote. Months passed before I knew whether Toama's family was Sunni or Shiite; adding yet another wrinkle was their habit of putting a simple tree in the corner of their sitting room to mark Christmas. Over breaks for coffee, punctuating the day, everyone knew to steer clear of politics, that badge of sectarian identity. A simple lament for Lebanon usually sufficed.

If it didn't, George Jaradi would assume the stage.

George had trouble walking, a problem I originally put down to either a long-ago injury or the bottle of arak that he was constantly hitting. Toama's cousin, George was officially a mason, but this was only a title. He was charged with building a stone wall to mark the border with Massoud's land, sandblasting the house's stone façade, and repointing the ninety-year-old mortar. It was a remarkable amount of work that he rushed through. George, never a slave to schedule, actually seemed determined to meet our deadline.

"We want movement! We want work!" George shouted one day.

"*Haraka* is *baraka*," I told him. Movement is blessed.

The words provided him a sense of triumph.

"*Yalla,* Abu Jean! *Yalla, shabab!*" he shouted. "Let's go! Let's go! George wants you guys to wipe your brow and see sweat on your hands."

George, who surely was once handsome, his tall and dark features evoking an Egyptian movie star, was weathered now, his face lined and sagging in the way of lifetime drinkers stoked for years on Marlboro Reds. Worn day in and day out, his sweater and pants never quite fit. Most noticeable, though, was his propensity to refer to himself in the third person. Then there was his walk. While working construction in Sidon during the civil war, he had fallen off the third story of a six-story building when a bomb detonated nearby, hurling him to the ground.

"It sent George flying," he told me. He broke his left leg, both wrists, and his back. "George was conscious. George never lost consciousness. George said, 'Don't move me! Don't move me!'"

Of course, no one listened, and none of the bones healed as they should have, but by now his limp was a signature gait, as he sauntered awkwardly, like a street-smart ne'er-do-well in Cairo, with a bit of Gene Kelly, managing to head somewhere fast without showing that he was in a hurry.

One day, after the weather had turned cold in November, I was standing downstairs with the increasingly indolent Abu Jean, trying to figure out what he was doing besides smoking, waiting for more coffee, and jabbing useless tips down to Toama. "George has been here since the morning!" George declared. "Where were you, Abu Jean?"

"I was here, too," he improvised, a bit more embarrassed than he ever got.

George was in rare form this day, and nothing was sacred, not even the coffee that arrived, yet again, in a red kettle atop a battered silver tray, with four small white cups.

"Fuck the coffee!" he shouted, three times in a row. "A little movement and we'd be done. There's no time for coffee today. There's only time for work." George saw me smiling at his theatrics. "The work's got to be like this!" he told me, thrusting his fist forward like a boxer's jab. "We've got to finish!"

Abu Jean smiled. Abu Aaajah, he had taken to calling George. Mr. Loudmouth might be the best translation.

George shook his head. "Abu Jean, George just wants movement!" he told him, words that were meant for me. "*Yalla ya shabab. Yalla!*" He said it twice more, followed by "*ya akhu al-sharmouta,*" that brother of a whore — apropos of nothing.

Throughout the morning, it was all theater and swagger, as if a parody of what work should be like at a *warshe*. For George, silence was only a reluctant pause to ponder his next string of expletives.

The next day, George's stone wall was rising behind the house row by row, a piece of thick fishing line stretched across it to keep the sum of its parts at the right angle. George, manic as usual, wore his uniform of gray knit cap, frowsy sweater, and white gloves encrusted with cement. A cigarette hung from his mouth, as it tended to most of the time.

Abu Jean's blocks, suggested George, were merely orderly, piled one on top of the other, but his own walls, intention evident, suggested a maze. George had found the stone, but then had to craft it, as all depended on the façade and tricky angles. "This is its face," he said, lecturing his apprentice, Haytham, and pointing to the front of the stone. The word was feminine, as was his meaning. It would look beautiful. "This is its ass," he said, turning the stone around. It took ten blows to shape everything but its ass, which no one would see. He tried to intimidate it, pummeling and bludgeoning, drubbing and flogging it. Next he reverted to persuasion, tapping and patting it, the equivalent of a caress.

"Brother of a whore!" he shouted. "It doesn't want to break."

It finally did, cracking along the wrong fissure, the latest confirmation, for George, of an angry universe. He threw the mutilated chips behind the wall with the pebbles, sand, and rocks. Then, always resilient, he began again with another lover — wrestling, cajoling, pleading, intimidating.

"*Shou*, Abu Jean, this, does it please you?" George demanded to know after a labyrinthine search for another stone.

Abu Jean muttered something.

"*Yalla, ya Toama!*" George shouted. Get a move on!

Toama, as was his habit, ignored him, and George started another search as, miffed by Abu Jean's dismissal, he roused himself for further debate. Abu Jean was responsible for acquiring the stones, and George, to needle him, diminished the lot with a hasty but impressive contempt.

"They're all good," Abu Jean countered.

George began kicking them as he walked down the path. "Fuck this! What's so good about this one?" He kicked another rock. "And this one?" On it went, down the line, repeated again and again. "Fuck them all!" he declared. "Toama," he said sternly, "your and Abu Jean's work today doesn't please George at all. Not at all."

With unmitigated assurance, one *maalim*, a carpenter, had told me that Isber Samara had built our house 112 years ago. *Absolutely.* Wrong, I countered. It was not yet a century old, I claimed, as he shook his head with the vigor of a desert hermit ordered to live in the city. The carpenter was so convincing, I wondered if I was wrong.

Nothing about the house was really certain. Most fact-gathering led to bewilderment, and once again I was overwhelmed by futility. At times, my mood turned dark and stormy. I had resumed smoking, returning to form with the acceptance of an existentialist and the gestures of a madman. On clearer days, I would try to grasp what sort of odd mishap we were perpetuating—an insult to architecture seemed likely. I slipped toward the abyss as, without my realizing it, the project sputtered toward something not finished but not in immediate danger of collapse.

George and Toama had cleaned every stone outside, along with the walls and arches downstairs. The stones themselves were revelatory. With the new mortar, the angles of the structure pulled things back to order and precision. Colors were wrestled from time, as shadows gave over to shades of cream infused with many hues bordering on blue. Downstairs, electricity had been restored, wires buried in rivulets of concrete running along the floor. Doors and windows came off as sunlight filtered into corners unknown since the house's infancy.

The house was a painting gradually emerging from my endlessly deliberated yet haphazard choices. Occasionally, I feared that we had too radically eviscerated the original, though later I would beg for alterations more transforming than any previously imagined.

I was hesitant, unsure, and equivocal in decision after decision, read by *maalimeen* like Abu Jean as weakness and vulnerability. Whenever Abu Jean sensed it, he became a taxi driver in Cairo, predatory in his knowledge. We walked through the house, and I suggested to him that we refashion into tables the old doors that were salvaged from downstairs and seemingly beyond repair. *Maybe,* I said.

"What kind of talk is this?" he shouted at me, the cigarette in his mouth rising and falling with the hectoring words. "People will start laughing at you."

As with Massoud Samara and the fig tree, I harked back to my ancestors.

"This is from the days of your grandmother. This is from the days of your grandfather," Abu Jean said sarcastically, pointing at the arches, windows, and marble, one by one. "God have mercy on your grandparents!" It was part joke, part respect, and part weariness from having heard my line too often.

The threat of war or violence is as common in Lebanon as the wind, and just as unpredictable. Isber Samara lived with it, although he seemed, for a time, as his house rose, to distance himself from it. As his family became accustomed to its splendor, he no longer could. The future was unresolved and full of danger. Around him and his family, Bedouins and gangs were pillaging the countryside. New boundaries, marked by stakes hammered into the ground, began to divide what had not been truly united but had never been separated. The French and British negotiated the future of their imperial possessions. Bab al-Tniah, Tal al-Nahas, and other new customs posts promised to sever Marjayoun from the Houran, an expanse that men like Isber seemed to need in order to breathe.

Two codes of justice, old imperial and new colonial, clashed and confused. Economies changed, currencies multiplied in the wake of the Ottoman Empire's collapse. First came the Egyptian pound, pegged to the British sterling, then the Syrian pound, fixed to the French franc. Sectarianism and nationalism, the dangerous kinds, reared their heads in spectacles of horror and cruelty. The Ottoman influence remained, but its old connections and routes buffeted the new borders and were redirected and skewed into something artificial and awkward, like broken legs miscast.

Isber knew the debates. He had heard his neighbors consider and name their loyalties — to Arab nationalists, to the French, to a Russia mired in revolutionary turmoil and ceding leadership of the Orthodox Christian world for the vanguard of the Communist International. He had heard of the American investigators of the King-Crane Commission, who in 1919 toured Palestine, Syria, Lebanon, and Anatolia. As France and Britain cynically divvied up the spoils of the Ottoman Empire, securing oil for their citizens, the United States dispatched the commission to ask, rather than tell, Arabs what form of government they wanted. Predictably, its findings were ignored. (It was said that Marjayounis favored Arab independence, then some kind of American control.)

Isber told his family that even if war and rebellion were prevented, Lebanon had no future. At least his Lebanon, the swath of territory stretching unbroken from Wadi al-Taym to the Golan Heights and into the Houran. The realization pained him.

For years he had prepared to send his oldest son, Nabeeh, to medical school, a step that would certify his family's entry into the elite of a town that revered education. But Isber carefully considered the education of all his children. Whenever schools were closed because of strife and violence in the countryside, he managed to bring tutors to the house. (They were much taken by the creativity and talent of his daughter Ratiba, who was already penning poetry in Arabic.) He worried about the children's educations, their futures. His Lebanon was perched before an abyss, more unpredictable than the Great War, and nothing — not France, not Arab leaders, not the British army across the border, not the potentates of the old order — could pull it back.

As 1920 arrived, Isber seemed to smoke more. He lost weight, his face becoming more drawn and his mustache, in turn, becoming more pronounced. He had spent his life trying to make his own name and achieve wealth and the sort of success admired by others. More recently, though, his goal had become securing a future where his children could realize their ambitions and create their own families without the distractions of fear and conflict. As the lawlessness around Isber grew, it was clear that the finest local schools, including the medical school, were not enough to ensure his children's safe passage into adulthood. He worried that Nabeeh, his oldest, might be shot by bandits, who killed indiscriminately. He worried that his two older daughters — Nabiha, then sixteen, and Raeefa, then twelve — might be murdered or raped if the violence continued, especially in the event that he and Bahija were killed or maimed.

America offered peace and opportunity, neither of which he expected to see again in the land where he had spent his life. The thought of losing his children to death may have touched an old wound that Isber Samara had carried since he was a young man. His older brother Said had worked with Isber and his other brothers at the beginning of their careers. Said was just twenty-three years old, yet to marry, when the brothers started out to create a business they could share. For months at a time they traveled in the Houran, often sleeping on the fine black dust of the steppe. One night, a horse, spooked by the cries of hyenas and jackals that still roam the Syrian wilderness, trampled and killed Said as he slept. Said was three years older than Isber, and Isber had carried the memory of his brother's loss ever since. This time, Isber was determined to find a way to cheat death. His children would live safely without constant fear of its approach.

All his children. All their lives.

On a cold November night, Shibil and I sat in chairs, leaning forward, ears to the radio, listening to a speech by Hassan Nasrallah, the Hezbollah leader. There was no electricity. The radio ran on batteries. The fluorescent light was plugged into the old car battery that he had yet to replace. The words that tumbled from the speaker were vintage Nasrallah, the most charismatic Shiite orator in a generation.

Israel's scheme in Lebanon is that of sedition and of encouraging the Lebanese to fight each other, the Hezbollah leader told a sprawling rally in Beirut. His words, defending the group's weapons, what it calls "the resistance," were stentorian, as he assumed pulpit and stage. Nasrallah had an innate sense of a crowd. He built an argument using formal Arabic vocabulary, then underlined a point in conversational idiom. His most emotional refrains were delivered bluntly, subdued by jokes, a few words in a softer voice, and sometimes a quick aside that his slight lisp did nothing to diminish.

His words echoed the landscape around us, the posters that Hezbollah had put up in parade-like intervals down the road from Shibil's house. *The Time of Defeats Has Passed*, read one variation. *The Time of Victories Has Arrived*, declared another.

Brothers and sisters, Nasrallah said, Israel talks about peace at a time when it beats the drums of war . . . No one should rule out the possibility that the Israelis intend, in cooperation with the Americans, to drag Syria into war and drag the entire region into war, for this was George Bush's plan in the region. The plan, Nasrallah declared, is not that of peace, but of war.

"It's pretty strong speech," I said to Shibil.

"Hmmm," he answered, nodding, in a tone that conveyed approval and worry.

• • •

I returned to Shibil's the next night, and sitting on the couch, I surveyed the room. Next to his chair was a trash can filled with the red shells of raw pistachios. On the floor was a copy of *Mad* magazine from September 1999. "16 Pages of Extra-Stupid Movie Spoofs!" it shouted. "18 Things to Do with a Live Lobster (Other Than Boil It Alive!)." He once assured me there was an eight-track tape of Deep Purple's *Made in Japan* tucked away somewhere.

As we sat in his house, I felt a wave of sadness as I watched him. Shibil — ailing from an arthritic hand and creaking joints — was suffering, deserted by his friends and family. Of his four brothers, only the oldest maintained a relationship with him. The others found Shibil an embarrassment, a failure in a family — like most in Marjayoun — that still prided itself on education and wealth. The brother closest to him in age sometimes did not even greet him when he visited Marjayoun, a rather remarkable feat, given that they stayed in the very same house (albeit with separate entrances).

It was not a matter that was kept inside the family, either. A few weeks before, I had joined Shibil and two of his brothers at a restaurant on the Litani River. It was a charming spot, a simply built patio nestled in orange and lemon groves on a verdant bend of the river, where children swam and fished in water that cascaded over shallow stones. The rapids skipped along on a wind that barreled down the valley, past the pine trees that represented the sole demarcation of a vanished palace. There was no better fish served anywhere — trout stuffed with garlic and lemon peel. The radishes were inexplicably delectable.

During the meal, neither brother much acknowledged Shibil, who never looked up from his glass of scotch. I wondered if he was conscious of the slight, or lost in some private screening of an American movie, trying to catch mistranslations in the Arabic subtitles, through the smoke of marijuana and inebriation of drink.

Hesitating, I asked him about it afterward.

Families always have conflict, he told me, reluctant to say too much. "From the outside it looks okay, but from inside the family, from inside the house, you don't know." He seemed to be hiding his pain; if he ignored it, it might hurt less. "In any family, there are always feuds.

They don't pass in peace all the time. It's a general rule, *yaani*. Could you find a family that's so peaceful no shout comes out? I doubt it." He stopped, and I thought it better not to press.

One night, I called Shibil to see whether he wanted to visit a friend, Assaad Maatouk, a former chef in America who had returned to Lebanon to find a wife, but he was still sleeping off a lunchtime scotch. I waited for a while and then, bored, went over to Shibil's anyway.

When I arrived he was in his usual lackadaisical mood, but the tragedy was that it had not always been so. After returning from Oklahoma for the last time, just months before the Israeli invasion of 1982, Shibil worked for four years teaching English at the prestigious Marjayoun National College. He was an *ustaz*, a teacher, and while it did not make him wealthy, the job itself demanded respect. Next he taught at the Greek Orthodox school in town, followed by five years at schools in Khiam, across a valley and a short drive away, before returning to the Orthodox school for three more years, until 2000.

"Ever since that day, I haven't taught," he said, never quite explaining why.

Yet he insisted that his former students address him as *ustaz*. They didn't. By now Shibil had lost their respect. Beyond memories, he no longer counted anyone as a friend. Day after day, Shibil rarely left the house other than to buy water; it was as if he was dying a slow death of the sort claiming this place: Having lost its mercantile prowess, the no longer thriving Marjayoun survived on the aid, remittances, and generosity of its expatriate children. As with Shibil, who saw no shame in accepting money, and even expected it, the town waited for charity, no longer expecting anything from itself.

"I'm fed up with the whole shit," Shibil told me.

A movie was playing in the background. He glanced down at his chair, sagging from the years and draped in a once colorful sheet now indistinguishable from the furniture. "This is shameful," he said, pointing to the bedraggled seat. "This is my ass. It eroded it."

He sank deeper into the chair, smoking more as the night wore on.

"This is fate. *Que sera, sera.*" We started talking idly about how to render the phrase in Arabic: *Allah yikhaleek,* God keep you; *Allah yi-*

hadi al-bal, God give you peace of mind; *Allah yitawil omrak,* God give you more years, and its equivalent, *Allah yaateek al-omr.* Then he mentioned *Bi-sahtak,* Take care of yourself.

Shibil tilted his head back, as if surrendering. In a few minutes his eyes were half closed. The movie played on, the sound turned low but audible as it washed over his face. Toward the end, one of the characters, bitter and hating himself, declared, "I'm a garbage man of the human condition."

9

MR. CHAYA APPEARS

I SUSPECT IT IS not at all surprising that my quest to find the most alluring elements of the house's return to grace would become, in my mind at least, a small odyssey. Would I have wanted it otherwise? Wasn't this the sort of meaningful search I had hoped for when I took on the project of revamping the old house? All I can say is that one destination led, almost magically, to the next, with the characters best able to ensure my satisfaction appearing right on time, as if someone, somewhere, had the inclination to guide me on my rather quixotic mission.

The Maalouf Trading Company in Beirut felt like a door to another epoch where haunting discoveries lingered amid a deep, rich silence. No street voices filtered in here and every utterance seemed an awed, respectful whisper. Maalouf's was a pocket of entrancing stillness and impressive dignity set apart from the loud, tumultuous world. The smells were antique: dust, mildew, and the salt of the sea, just a short walk away. The Levantine tile of old houses spilled across the sloping floors. Vessels and artifacts stood before me in quiet repose. Old and delicate, they seemed beyond contact and certainly far from utility. They waited in peace, reminders of what seemed less troubled times but perhaps were not.

After the fall of the Ottomans, Britain and France remained entwined in the Middle East. The former had made promises, freely but casually, to the family of Feisal and to its rivals in Arabia, to Zionists bent on founding a Jewish state in Palestine, and to the French themselves. The two imperial powers were bound by the Sykes-Picot Agreement of 1916, which staked each country's claim to Ottoman spoils after the war ended. There were too many promises, too few specific, considered intentions. With a sense of inevitability, the French and British reached a deal in Italy in 1920, and the French were soon lording over their new domain.

The government of Mourad Gholmia, partisan of Feisal, fell two months after it was proclaimed in Marjayoun, when two hundred French horsemen rode in to take control. Entering the Serail, they removed Gholmia's hastily stitched flag. Feisal's government was crushed at the battle of Maysalun on July 24, 1920. In a story apocryphal but telling, the French commander, General Henri Gouraud, rode into Feisal's capital the next day, proceeded to the tomb of Saladin, and kicked it. With words obviously chosen for history, Gouraud declared, "Awake, Saladin, we have returned. My presence here consecrates the victory of the Cross over the Crescent."

France soon created a country where none existed. Lands once joined by history, tradition, clan, and commerce were divided by imperial borders put forth by the loudest voices. In Lebanon, these were the Maronite Catholics, a Christian sect long united with the Roman Catholic Church. Ties between France and the Maronites stretched far back, and in more modern times France had played guardian to the Maronites, who had enjoyed a certain autonomy under the Ottomans. The Maronites' leadership — religious and political — had long pushed for enlarging their homeland, and after the Ottomans' fall they saw the opportunity to create a greater state. The French, albeit with reservations, complied. To the Maronites' relatively small sanjak, or district, the French added the

coastal towns of Tripoli, Beirut, Sidon, and Tyre, all of which had be-
longed to the Ottoman province of Beirut. From the Ottoman province of
Damascus, they added the fertile Bekaa Valley, which included territory
that encompassed Marjayoun.

The Maalouf Trading Company was a vestige of Isber Samara's age.
Composed of what Americans might call a showroom, filled with unexpected alcoves and secret spaces — the family villa, a warehouse, and
other unseen sections — the building hid itself in the shadows around
the corner from the Café of Glass, a defiantly unfashionable locale set
amid the booming nightlife of the neighborhood. Facing Maalouf's
was a store lacking its mysteries, a place seemingly attempting to compensate for its quotidian presence with aromas of cumin, pepper, and
cinnamon, which scented the entire street in this venerable quarter of
Beirut.

At Maalouf's, one room flowed into the next, beginning with the
chamber just beyond the entrance, with its stone vases and urns, stacks
of tiles, and exquisite lamps. The walls were unpainted, pale as fallen
leaves transformed by time. Crumbling, peeling, and falling, the plaster concealed dark stone underneath.

Taciturn, but given to wry attempts at smiles that only slightly
qualify as such, Michel Maalouf was a trader of proud expertise, and
anyone who challenged his confidence had surely failed to take in the
import of the store's vast, impressive inventory — the harvest of pillaged houses and a lost Levant. The display of his goods was compa-
rable to the antiquities on exhibit at the Egyptian Museum in Cairo,
where sarcophagi and scarabs emptied themselves amidst the clutter
of history. Doors were stacked twelve deep, propped against a wall,
some engraved with grapevines snaking together. Others, embossed
with the traditional three squares, like the oldest doors at Isber's
house, revealed their touches of artistry with great reluctance. Everything spoke of subtlety. Stones, decorative in intention, crowded the
floor and ornamented gates. Windows, vases, and urns were offered in
every shape — squares, rectangles, circles, and ovals. Calligraphy com-

plemented the designs. Toward the back I glimpsed wooden tracery pulled from the triple arcades of residences once exquisite but now almost certainly demolished, faded into memory. As many as a hundred pieces of tracery were stacked on top of each other, their slender forms and fine woods too tender and fragile to withstand customers less than completely careful.

Only when I had become hostage to the place and lost sight of my actual mission did I finally make it to the courtyard for a close inspection of the cemento tiles. Laid before me were pieces containing stories and scenes, distilled images, and moody skies of places and eras so long gone that the words of any witness to them had faded before time. There they all were, in the autumn sun, wrapped in dusty plastic, stacked on wooden pallets, waiting. I wanted to use them in Isber's house but was anxious about the expense.

"Twenty-five dollars a square meter," Maalouf told me with firm assurance.

That translated to a dollar a tile, a tidy sum, considering that even a modest room of the house in Marjayoun required hundreds of squares. An entire floor of the house would require thousands of tiles. Maalouf acknowledged my shock with no more than the suggestion of a shrugged shoulder.

"What is it you were thinking?" his eyes seemed to say. A bargain, clearly, was not the intention; such would be neither appropriate nor respectful. It would, in fact, be an insult.

No one is absolutely certain of the origin of the tiles that I coveted at Maalouf's store. The mystery of the cemento's past added to the allure; no place or time could claim it. Some say Italian craftsmen, inspired by the natural colors of granite and marble dust, were the creators. Others claim the tiles were first produced in France, at Viviers, at the edge of the Rhone, around 1850, alongside the first cementworks there. Workshops, it is said, soon became common from Lyon to Marseilles. But not everyone believes this theory. Some say that the tile was first mentioned in Barcelona in 1857, and that from there, it flourished throughout Spain.

Before World War I, the tile was considered the epitome of what was splendid, and covered the floors of tsars' palaces, mansions of the Côte d'Azur, and offices of great import in Berlin. Cemento is also present in Barcelona, in the work of the architect Antonio Gaudí, who designed some tiles himself. Colonists took the tile to the imperial possessions of France, Spain, and Portugal. Along with its high commissioners and chasseurs, France brought the tile to the countries of the Levant; it became so commonly associated with things Levantine that it seemed to stand for that lost era. Before its fall from fashion in the 1950s, when cheaper, drearier alternatives arose, cemento appeared in the Philippines, Brazil, Chile, and Algeria.

In the Middle East, the tiles came to be known as *sajjadeh,* one of the Arabic words for carpet. But there were myriad other labels: Cuban tile, *mosaicos hidraulicos,* mission tile, encaustic tile, hydraulic tile, *carreaux de ciment,* redondo tile, *ladrilhos hidraulicos, impasta,* Barcelona tile, Portuguese tegels, *rusticos,* and so on. The language was the metropole, from where the tile emanated. The designations were the periphery, the far-flung locales to where they were destined.

The designs ranged from utter simplicity—a pattern of lines in black or gray—to three, four, or more colors arranged in intricate geometric, stylized floral, or other patterns. Rarely, animals or human figures were represented. Styles spanned cultures and times, blending and fusing: Influences included art deco, art nouveau, neo-Islamic, and so on. Some patterns defied category; some especially unusual ones seemed like watercolors.

The tile was as beautiful as it was simple, hence its revolutionary aesthetic. Rendered by hand, no two tiles could be exactly alike. In a sprawling room, hundreds piled together, their numbers multiplying, each had its own character and each told a story. Or conveyed a mood or ambiance. Most meaningfully, they crossed borders that, for a time, were still crossable.

Again and again, as I had wandered through Isber's house, my mind had journeyed back to the past. When I gazed at the arches buried beneath cement for decades, I envisaged the stately entrance created nearly a century before. I imagined Bahija polishing the tile and marble on her knees until it reflected her eyes, perhaps sad, perhaps concerned, as they considered the fate of the land now called Lebanon.

On September 1, 1920, the French high commissioner, General Gouraud, proclaimed from the porch of his official residence in Beirut the birth of the state of Greater Lebanon. Soon the French would draw further arbitrary borders to divide their newly won domains in neighboring Syria. By the time the colonial powers finished, a territory previously divided only by the geography of mountains, rivers, and valleys was transformed into a puzzle of political divisions represented by five states of uncertain identity: Palestine, Lebanon, Iraq, Syria, and Transjordan.

Once a gateway to Damascus and Jerusalem, Sidon and Haifa, Marjayoun was now part of Lebanon, pitifully small and dominated by Maronite Catholics, aligned with the French. Its environs had been incorporated into other new nations. Parts of the Houran were, for a time, two borders away. The new frontier between Lebanon and Palestine was eventually demarcated by a British and a French officer on horseback, who put a stake in the ground every two kilometers. Lost was the Hula Valley, an expanse long frequented by Marjayoun's traders and landlords. Its rich farmland, with its lake and marshes, was now included in Palestine. (In 1947, the United Nations made it part of Israel. Marjayounis were never compensated for properties seized.)

Customs posts traced borders that defined the reach of the French in Syria and Lebanon, the British in Palestine. Trade routes were severed, landholdings partitioned. Towns like Quneitra, Haifa, and Jerusalem, where residents of Marjayoun had worked and visited for decades, became more distant.

Isber and the other Greek Orthodox in Marjayoun were more comfortable as Arabs—not only in language, but in customs, tradition, and history, still drawn from the Houran, where some Christians continued to live as Bedouin tribes. To some, they were more Arab than their Muslim neighbors. And, after World War I, memories of Mourad Gholmia's hastily stitched flag remained resonant. Sympathy ran deep for Feisal and his ambitions for a greater Arab kingdom. All this irked the Maro-

nites, who chanted at Marjayoun's residents: O girl in the red skirt with the ruffled trim, the French are now governing, so let the anger kill you, O Orthodox!

The French divided as they conquered, favoring Catholics over Orthodox, Christians over Muslims, the countryside over the city, and minorities over majorities. While many Orthodox may have been skeptical of Maronites speaking on their behalf, Muslims were aghast at the notion of being ruled by a French overlord, a Christian one at that. Opposition among Muslims percolated — from gangs exploiting the prevailing lawlessness to insurgent bands fired by nationalism. Sectarian agitation had begun to rumble across the country. As each year passed, it deepened dangerously. One chronicler called it rare to see a man lacking either a French- or a German-made gun.

Subtle and coy, the cemento at Maalouf's did not speak of war, or frontiers, and the spaces they narrowed, but, rather, grandeur. The tiles returned one to a realm where imagination, artistry, and craftsmanship were not only appreciated but given free rein, where what was unique and striking, or small and perfect, or wrought with care was desired, where gazed-upon objects were the products of peaceful hearts, hands long practiced and trained. War ends the values and traditions that produce such treasures. Nothing is maintained. Cultures that may seem as durable as stone can break like glass, leaving all the things that held them together unattended. I believe that the craftsman, the artist, the cook, and the silversmith are peacemakers. They instill grace; they lull the world to calm.

The tiles at my feet were the remnants, in Arabic the *atlal,* of a lost Marjayoun. They were artifacts of an ideal, meant to remind and inspire, vestiges of the irretrievable Levant, a word that, to many, calls to mind an older, more tolerant, more indulgent Middle East. The Levant was, in part, a geographic concept. Loosely defined, it stretched across the eastern Mediterranean, the arc of the Fertile Crescent, the frontiers at the Isthmus of Suez in the south, and the Taurus Mountains in the north. But the Levant was really more a culture than an expanse of land or group of nations or homelands. It was a way of living and thinking that bound Asia Minor to the Middle East and Egypt to Mesopotamia. It was, in essence, an amalgamation of diversities where many mingled, a realm of intersections, a crossroads of language, culture, religions, and traditions. All were welcome to pass through the territories and homelands within its landscape, where differences were often celebrated. In idea at least, the Levant was open-minded, cosmopolitan; it did not concern itself with particularities or narrow definitions or identities. Isber Samara, whose house was a Levantine expression, could roam from Hayy al-Serail to the volcanic plains of southern Syria, past the wildflowers and basalt cliffs of Mount Hermon's snow-capped range. In the time of the Levant there was freedom to savor the worlds of others. But borders, rendered with caprice, ended what had been. Cemento was not just a relic of the time; it was a tribute to the imaginations it nourished.

This is why I wanted it in my home, but Maalouf was a tough negotiator.

"Twenty-five dollars," he repeated once more, no less certain than before, as we kept haggling in his store over the price of tile that had become my obsession.

I shook my head, and he pointed to the wall, against which a vast ceiling panel leaned. Obviously wrought by loving hands, the panel was composed of twenty-two squares of eight designs — a purple flower, an arabesque vine of blue and red blossoms, an inverted fleur-de-lis painted the same green as the doors, and more somber plants, carved but unpainted.

"Uff," he grunted when I estimated its worth. The piece was going for $12,000.

The rest of his rooms were filled with more of the same: There was *darabzin,* iron railing finished in black or gold. A carved marble fireplace was propped against another wall. A testament to the divine right of a monarch, the hearth appeared more European than Arab, more emperor than caliph. Two columns towered over everything. One was inscribed *In the name of God.* Their capitals, no longer attached, stood with no burden.

In these negotiations with Maalouf, I tried to be Lebanese, attempted to gain leverage, sinking to a new low: manipulation of shipping charges. I talked fast, tried not to give up, went back and back, attempting to shave off pennies of the cost of transporting the tiles to Marjayoun. I suggested the quality was, well, not so great and made excuses — my price had been put forth by the project's engineer, foreman, my architect, and the tile layer. Experts all. This is what they said was not fair but generous. Maalouf simply ignored me.

"What's twenty-five dollars, really?" he said, as if my paltry concerns were irrelevant.

The instructions of the man named Abu Ali were clear. Wait outside, don't enter the parking lot, I'll find you, he said by telephone. The taxi dropped me off on a street in Beirut, and fifteen minutes later, Abu Ali showed up in his dark green Toyota 4Runner, pulling up at the curb near a McDonald's and a Spinneys, a British supermarket chain.

Slight and haggard, but with a hint of menace, Abu Ali was smoking a Marlboro Red as I climbed into the front seat. We exchanged a few pleasantries before he took a phone call, berating someone. "Don't do it again," he instructed icily. Abu Ali spoke slowly, clearly, in tough little riffs.

The negotiation that ensued for more cemento had a surreptitious tone. Meaning, lots of vague pronouns.

How much did I want? Abu Ali asked me, voice low.

As much as he had, I told him.

He told me he thought he had thirty-five or so square meters.

No broken pieces, I emphasized.

What are you going to do with broken pieces? he asked me.

I suggested paying half now, half later, and he seemed shocked. It was all so practiced, all so smooth, yet a little clandestine. I felt as if I was buying heroin. Any minute the sirens would start flashing.

Merchant, connoisseur, looter, Abu Ali had no office; he made deals in the midst of the rubble he created as he tore down historic buildings. Abu Ali lived in a hardscrabble neighborhood, a quarter that remains among the most political of Lebanon's urban geography. The

Dahiya, it was called. Over it flew the yellow banner of Hezbollah, its color taken from the breastplate of Imam Ali. To Hezbollah's supporters it was terra sancta, the bastion of their steadfastness.

Abu Ali's real name was Hussein Ali al-Bureidi, and like many in the Dahiya, he had originally hailed from the Bekaa Valley, a town near Baalbeck famous for its thieves, stone, and fruit. His parents had come to Beirut looking for work, and brought him, along with his six brothers and four sisters. Abu Ali, twelve then, dropped out of school in a few years, became a carpenter, then found his calling, becoming a merchant of destruction — a deconstructor — and a salesman of the things this activity unearthed. Someone had to tear down what remained after the soldiers departed and the guerrillas took off.

Abu Ali's work was profitable only because of the valuable leftovers he took: oak doors, window grills and ironwork from balconies, cemento and other tile, and anything portable in steel or stone. Sometimes he supplied builders with material. His estimate: seventy percent of the houses restored in Lebanon bore his mark.

"People have come in here cursing me: 'What are you doing taking all this old stuff?' Some people, friends even, get disturbed by the fact that I'm tearing down these old houses. But I love old houses more than they do. I tell them, 'If I don't do this, someone else will.'

"There are many, many old houses. Always will be. I've been at this for twenty years, demolishing houses, and they keep popping up," he said.

Abu Ali had two basic cemento designs, sphere or diamond. The colors were earthy: green, gold, and a remarkable purple that threatened to become brown. For years they had been secured to the floors of a three-story villa, one of the last historic houses in the neighborhood of Museitbeh, along the sprawling road that leads to the airport and, eventually, Marjayoun.

Thirteen dollars, Abu Ali offered as a price for each square meter of the tile. It was almost half what Maalouf demanded. Don't act surprised, I told myself, shifting to a moderately convincing impression of disinterest. I had to think about it, I said.

And so we returned to Abu Ali's Toyota 4Runner, pulled up near that McDonald's, where the negotiation would end. He never put himself in the position of begging. He was the giver, the one helping. It all reminded me of a line my landlord in Beirut had once told me. As we haggled over my rent, he declared, with a hint of irritation, "I'm not going to cheat you." He paused, reconsidering his words. "Well, I might cheat you, but not that much."

Abu Ali knew he had the deal before I got in the car. He started at $455, about $13 a square meter. I countered with $12. He agreed, adding another $100 for transportation, making the total $520. I said $500, and he agreed again, without hesitating. Without a pause. It was a clear sign of what I grimly expected from the outset. Abu Ali may have been cheaper than Maalouf. But I was overmatched, and got a bad deal.

If charm were indicated by a range of colors, Edgard Iskander Tannous Andraous Chaya was the deepest of reds. He was my last stop for the rest of the tile, and minutes after meeting the Lebanese merchant, I knew I would never address him as Edgard, sir, *ustaz*, or *monsieur*. I would always call him Mr. Chaya. Seventy-nine years old, Mr. Chaya was impeccably dressed, a silk scarf stuffed in the breast pocket of his suit. His pipe was a constant, though he didn't seem driven to reach for it, as I did for my cigarettes. When he did light up, the flame flickered past his groomed mustache toward the lenses of his square-rimmed glasses, and he smiled, as if he believed no one else could pull off so deft a gesture.

Gazing at a shapely Lebanese colleague who had joined me, he flirted zestfully. "You're simply a mannequin," he declared in appreciation, somehow lacking in crassness.

I took longer to assess. We began speaking in English, before he arbitrarily interrupted, as a look of surprise crossed his face. "Hold on!" he said. "I have clients from Bayt Shadid in Canada. Where are you from?"

Marjayoun, I told him.

"Don't you speak Arabic?"

"*Walaw,*" I answered. It took this expression — only one word — for Mr. Chaya to celebrate my accent. It meant something. "How may I help you?" he asked in Arabic, without pause. "You need tile for your house in the United States?"

"No, no," I said, "it's for *bayt Sitti*," my grandmother's house.

"For what?" he asked exuberantly.

"For *bayt Sitti*," I said again. "In Marjayoun."

"*Bayt Sitti!*" he cried, exaggerating my accent. "*Bayt Sit-tee!*"

To everyone there, he asked me to repeat the phrase, again and again.

Bayt Sitti! Bayt Sitti!

From our first meeting, Mr. Chaya was never anyone but Mr. Chaya. It did not matter whether I was addressing him or mentioning his name to friends. Even in my phone book, I wrote "Mr. Chaya," with reverence and respect. And from that moment, Isber Samara's house was, in Mr. Chaya's recollection, never again Bayt Samara. Nor was it my house or my family's house. It was *bayt Sitti*. As in: *Bayt Sit-tee!* As in, grammar aside, his voice rising in ecstatic celebration: *How is bayt Sit-tee?*

Mr. Chaya's family had a century of experience in cemento, but were less assertive than, say, the Maaloufs about reminding others of their extended experience. His grandfather had begun making tile in 1881, passing on the trade to his son, Mr. Chaya's uncle, then *his* son, Mr. Chaya's cousin. By 1975, though, the business was bankrupt, the victim of war, competition from cheap machine-poured tiles, and his cousin's lack of dedication. As a younger man, Mr. Chaya never believed that his family's pursuits could ever hold his interest. He became a moneychanger, growing wealthy. In 1995, at sixty-six and without financial concerns, he decided to retire and indulge his passion, sailing the turquoise waters off Lebanon's Mediterranean coast. It was not enough. "I got tired, restless," he admitted.

Because of his family, an elderly cousin gave him a box packed with brass molds. Rickety pieces resembling cookie cutters, they had not been used in a generation. "This was very important to your uncle," the cousin told him, as Mr. Chaya started playing with them — "like a child plays with Legos," he recalled. His interest was soon serious. No surprise, really. He had never been a dabbler in anything; he was always committed to achieving the finest. That was what he survived for. "It took me three years to produce my first tile. Imagine the determination."

He surprised himself: "It took me three years!"

He said it unlocked memories of his grandfather and his childhood. "I didn't come back to tile because I wanted to make money," he said. "I came back because of my roots. I like going back to the old, to the roots of things. My love for the old grew in proportion to my hate for the new and the modern." In tile, he felt he had discovered himself, and he was determined to make sure that the *atlal,* the remnants, of yesterday were never lost.

The civil war that began in 1975 extinguished an older Beirut. When the late Rafik Hariri, the tycoon turned politician who proved that no fortune was big enough, razed the warren of souks, Ottoman houses, and serpentine alleys that were what was left of downtown Beirut, Mr. Chaya bought twelve thousand tiles that had been salvaged. He purchased them merely to preserve them, before they were lost in the new downtown hub that Hariri envisioned, of half-million-dollar apartments and designer stores, faithful in style to its ancestor but shorn of any soulfulness beyond the frivolity of a playground.

"I hate things that are superficial and factitious," Mr. Chaya told me, smiling.

There was nothing imitative about the tile he had bought from Hariri. None approached the generic. Each had its character, its own mark, its distinguishing feature or quirk. "Sometimes, when I see a tile with a mistake, I laugh. It makes me happy," he said. "Because I know where the maker messed up and why. He was exhausted, he was sweating. And this is the beauty in making tile. It's human."

Eleven years later, his own tiles, inspired by the designs of the others, were gracing restored villas and the American University of Beirut in Lebanon, shopping malls in the Persian Gulf, a hotel in Ghana, resorts along Jordan's Dead Sea, and eclectic homes in Europe and America. He went from the smaller molds inherited from his grandfather to larger tiles, some twice as big. Each of his own designs included the name of a Beirut neighborhood or Lebanese village, from Hamra to Rawshe, from Sidon to Zahle. With three workers in his plant, he had produced twenty thousand or so tiles the year before, between twenty and sixty a day. Patience, he noted: Each tile took fifteen minutes to pour, one day to dry, and two weeks to cure.

Mr. Chaya's prices were high. Some designs, I would eventually learn, were twelve times more expensive than their century-old equivalents. I say eventually, because it took time, a lot of time, to learn this from Mr. Chaya, who had none of Abu Ali's desire to close the deal, none of an impatient Maalouf's appreciation of time. He could wait. While others fretted, he could appreciate the colors that filtered through the sky at dusk.

He talked amiably about acquaintances in the field, all those designers, engineers, and architects whose paths had crossed his. "I don't know them all, but they all know me," he said. "Like they say, *Tout le monde connaît de Gaulle, mais de Gaulle ne connaît pas tout le monde.* The world knows de Gaulle, but de Gaulle doesn't know the world."

On and on, Mr. Chaya chatted with me, devoting perhaps a tenth of the conversation to business. When I finally began to bargain, he charmingly but emphatically refused to do so. The tile would cost $25 a square meter, $1 a tile, the same price as Maalouf's most expensive version, even though Mr. Chaya's was new, absent of design, and relatively easy to manufacture. I was shameless, direct: "I'm spending so much on this house, Mr. Chaya. *Please.*" He would have none of it. His men were taking double in salary, fuel costs up, the sand so expensive.

I begged him. For the sake of *bayt Sitti,* I said.

"I don't want to refuse, but I can't," he said, never giving a cent. "*Wa hayat Allah,* believe me. I can't lose money. Shall I show you a breakdown of the prices?" He reached for his folder.

I shook my head.

He relit his pipe and exhaled slowly through his smile, leaning back in his chair.

How do you spell Edgard? I asked as I wrote the check.

When I returned to Marjayoun a few days later, the tile was already there, transported on flatbed trucks, three shipments in all, and stacked against the stone wall that George Jaradi was building in perfect order. Maalouf's medley of colors and patterns was mixed with Mr. Chaya's uniformly, pleasingly brownish style. Then came Abu Ali's thirty-eight square meters, three more than I bought. Abu Ali never lied. That day, and for more that followed, I gleefully, frenetically lost myself in the tile, as I once had with stories in Beirut, Baghdad, and Cairo.

With a red bucket filled with rainwater, a green brush, and a sponge so laden with grease that I was about to throw it away, I went through every single piece, separated them into their patterns, and each pattern into its respective colors. I created my own categories: most common (the pattern that enchanted me at Abu Ali's), most beautiful (arabesque wheels of a dozen spokes, in reds, greens, blues, and creams), and most graceful (an eight-pointed star alternatively cast in red and blue or brown and cream). On a free swath of concrete, I arranged a few dozen tiles into a patch of color, with a border forming a square around it. I brushed the tiles, washed them with casual swipes that always revealed something interesting — unexpected colors or swirls that at first seemed random, but when put together revealed some intricate design.

Within days, more stacks spread across the garden, in perfect piles of ten, claiming space between the shingles and a blackened steel barrel. I had counted more than two thousand tiles, but the time passed quickly. My hands were cut and dirty, and my back hurt. My feet were hopelessly muddy. George saw me, shaking his head in fulsome disapproval. "Anthony, you're going to make your hands rough," he shouted. "Leave them tender!" I smiled, then surveyed the scene. It felt as though I was lifting history and putting it back in its place.

In the days that followed, I measured each room in the house, twice. I walked the floor, abstracted, in a trance fit for an ecstatic Sufi, trying to imagine the tile that would soon unfurl across it. I designated numbers for the tiles and photographed them three or four times, inscribing each picture with a Sharpie. I sketched corresponding maps, the measurements calculated to the second decimal point. In lists, I accounted for each tile, giving them names — *red hex, red star, patio, Cave, brown star,* and *Abu Ali's star.* I scribbled diagrams on the pages of a reporter's notebook, making my best guess at how to render an arch on paper and the approximate — exceedingly approximate — place of doors and windows between the lines of a page. I guessed at how a meter of tile would translate to a centimeter of paper.

Early on, I knew I would never have enough tile to imitate the rooms of a Barcelona house or a Beirut villa, cemento from wall to wall. Of the tile I bought, nearly every batch was missing some of its border or one of its corners. Plenty were too damaged to lay — chipped, gouged, chafed at the edges, or worn smooth by foot after foot that had walked over them. I would have to settle for less, placing the tiles of Abu Ali and Maalouf in carpet-size spaces in nearly every room, surrounded by Mr. Chaya's plainer variety. Like the stone, the design of this so-called *bahra* was a necessity, a utilitarian use of too little tile. Like the stone, I hoped it would prove elegant as the house grew older. In the end, I envisioned Persian rugs spilling across the floor of Bayt Samara, muffling time.

I had a vision, flawed and impetuous, and like a Shadid, I wanted to fight for every chaotic angle that I had sketched on that pad.

· 10 ·

LAST WHISPERS

ABU JEAN HAD called someone to lay the tile, and as we stood outside the house, the bane of my mornings pulled into the driveway. I felt as if someone had thrown a power drill into my bathtub. There before me was the red Renault whose engine I heard being gunned at dawn, day after day, as I lay in bed. Suddenly awakened on endless and ever-colder mornings, I had imagined the driver. Now here he was. He had come to work here.

Labib Haddad plodded toward us, offering his large hand as Abu Jean took care of the formalities, the idle chatter that in Marjayoun had to precede any conversation. Then, after the delay for the sake of manners, came the business at hand. Could Labib lay the cemento piled on the wood pallets outside? The answer was affirmative, but, as usual, the timing was hardly definite.

Labib was one of the few tile layers in town. His occupation brought him attention. Few others could do what he did, with marble, ceramic, or the rarer cemento. But he left little impression on anyone; he rarely spoke, and when he did, his words were often so mumbled as to be indecipherable. That did have its advantages: In the conversation on that first morning and the ones that followed, Labib never said no to me. He never refused to lay my tile. He never asked for more time to finish his job at another *warshe*. He never asked for understanding.

Yet, in the days to come, he postponed and procrastinated, delayed and dragged his feet, literally. And it was his particular habit to do it all with a maddening shake of his head — as in, *This won't do*. I pleaded. To shut me up, he promised to come the next day.

He didn't. "I waited until four P.M. and the tiler didn't come," Abu Jean told me.

"Can you call him, Abu Jean?" I asked.

He started yelling — where is his phone number, whose phone should I use, what if he's busy? "If this person doesn't come, if that person doesn't come, what am I supposed to do? Should I pull them by their ear, drag them, and make them work?"

That's just what I thought he should do.

"Should I ask God to invite him over? Should I bring God from heaven and make him bring this guy here?"

I shook my head. The tirade had started; the gates were breached; the fleet had left the harbor. Only God in heaven was going to restore harmony, so I looked for my phone. After a few calls, I located our man, then got in my car and drove to a church in Khirbe — a neighboring town whose name means ruins, and was hence renamed, by its image-conscious residents, Borj al-Muluk, or Tower of Kings.

Labib was working there in the basement of the church, whose meek priest, I later learned, had driven the previous night all the way to Labib's house to berate him for taking so long to finish the floor. With me, Labib was friendly, with no sense of shame or remorse. He would come tomorrow, he told me again, nodding his head in assurance.

On an overcast day soon after, he did show up. Abu Jean had brought the cement. A friend of Toama's came to help lay the tile. A truck had dumped a load of sand in the driveway.

Labib wandered over to the tile. He ran his hand over the patterns, a finger along the edges. "This won't do," he told me, in his best imitation of regret. Before he could begin, each piece of tile would have to be cleaned and polished, he said. As I stood there dumbfounded, I could have sworn I saw a slight smile unfold across his face. "I would have told you if I had known this was the tile you were using," he said. Labib started walking away, waddling, crouched and hunched over, his back bent from tiling year after year.

"I'm around," he said.

So appropriate, I thought. He was going to wait for me.

Abu Jean, more than any other person, appreciated the scale of my disappointment.

Fittingly, the sky was lonesome and forlorn, with clouds blowing toward Mount Hermon as they turned dark and rainy. The weather was as mercurial as Labib's intentions. Rain did come, but gave way to drizzle. Wind sounded like a whisper, then a howl, as it poured through the house. I hoped I wasn't the only one hearing this.

"Let's go down to my house and have a cup of coffee," Abu Jean said. He made this suggestion nearly every day, sometimes two or three times. And nearly every day I politely declined. He changed his pitch this time: cactus fruit, grapes, apples, *and* coffee. Seeing that he was sensitive to my slight, I guiltily accepted.

His house was modest, yet blessed with a spectacular view of the valley and Mount Hermon. As we parked, he told me about his grandfather's sale of the family land for arak, the anise-flavored liquor that is celebrated and cursed. All that was left was this tiny plot, with its beautiful peach tree planted at the bottom. He said he tried to buy a small piece of land next to his, but the owner wanted far more than he could afford.

"Money," Abu Jean blurted out. "Curse its religion, that son of a whore."

We ate the cactus fruit, and Abu Jean lamented fate. Gone, he said, were the fig trees that once carpeted the valley, lush and green. When the people left, the figs began to disappear. Olive trees took their place — less demanding and more gentle — but they, too, were ignored. Olives were trees of peace, and there was no peace here.

"Twelve," Abu Jean kept repeating, growing angry and wagging his finger.

Quickly I learned that Abu Jean was twelve when he had to flee to Nabatiyeh, during World War II. Twelve years old. "That's war," he said. "We've had a hundred years of war here, son of a whore." The number meant nothing, but Abu Jean had become attached to it. It felt properly grand, even epic. "A hundred years of war," he repeated.

It was January 1920, and the rains had begun. At higher altitudes, along the mountain's peak, the rain turned to snow, which tumbled down the slopes of Mount Hermon.

Isber sat back in a chair that had come from Syria, bearing its damascene design. Another piece from the set bore the shades of wood — walnut, apricot, rose, olive, and lemon trees rendered in different shapes and sizes. Most of it was walnut, crafted in a technique that stretched back a thousand years, its mother-of-pearl from the sea, bahri, *and from the river,* nahri, *inlaid then accented with ivory. When Nabeeh came into the room, father and son sat under a portrait of Archbishop Elia Diab, which Diab himself had given them before leaving the town to minister in Chile.* Remember me *was written at the bottom in Arabic.*

There, among cushions that Bahija had stitched, Isber fretted. He wanted his eldest son to make his own decision, but behind those eyes that never revealed anything, there had to be emotion. Was his son going to say that he wanted to leave home? Would Nabeeh be the first of his children to say farewell, the beginning of his family's breakup?

The weather was cold, gloomy, and gray. The marble floor was frigid to the touch. Even in winter, Bahija could not bear to throw rugs over it, depriving the room of its glow, at its softest when the sun was at its most retiring. Such houses drew their heat from an ujaa *burning wood and sometimes* jift, *the crushed pits of olives they had pressed in October. Here, it was charcoal that glowed in the evenings.*

Now nineteen, Nabeeh had not returned to the Protestant school since World War I ended. Everything seemed too precarious for medical school, especially in faraway Beirut. Outlaws and rebels plied routes from the Chouf Mountains toward Hasbaya, Marjayoun, and on to the Houran. Predictable bloodshed began a series of raids and reprisals, vendettas and score-settling, flickering across the landscape for months. The worst episode erupted in Ain Ebel, a Maronite and Greek Catholic village far from Marjayoun but still in the south, along the present-day border of Israel. Residents of neighboring Shiite villages — along with the armed bands growing angrier at the French — had heard its townspeople were collaborating. Rumors spread that the French were arming them. Other stories followed: A rock was set up, declared the Prophet Mohammed, and used as target practice; Christians harassed a woman selling yogurt and tore off her head scarf. They soon attacked again, killing dozens, among them women and children, and burning the town.

The French, predictably, exacted revenge. All the strife and fear was too much. Isber wanted his oldest son to go abroad. It was the trek the family had made before — in their own myths, from Yemen, and in memories still tethered to their family names, from the Houran. This was another exodus, prompted for the same reasons, perhaps. But Isber was a father, and a father could not tell his son, especially his eldest, to leave, with the knowledge that he might never come back.

Nabeeh would have to decide.

Isber asked the question simply. "What would you like to do?"

Nabeeh had made his decision. Leave.

To Isber, there were two possible destinations: Brazil, where his nephew, the son of his brother Rashid, had gone. Or America, where his other relatives sought a future.

Nabeeh chose America. He knew of a man from the Fayad family who was going back. Maybe twenty others were going with him, Nabeeh said.

Never one to express emotion and never one to speak too much, Isber simply nodded. Then he made a request. Take Nabiha, his oldest daughter. Nabeeh agreed. What about Raeefa? Isber then asked. She was his second-oldest daughter, maturing herself.

"Let's wait," Nabeeh said, "and not take two girls with me at the same time."

• • •

The whispers of Isber's age were everywhere in Marjayoun. They spoke of a time in which I never lived but had envisioned so often that, to me, it was almost more familiar than the present. It was the era whose fragments and civilities — the remnants of the Ottomans and the Levant — had originally drawn me to the Middle East. I often wondered whether Isber, whose life was altered so unexpectedly by the events of his era, felt as dispirited as I did about the fate of his homeland. How deeply had he felt the loss that led me to his house? Undoubtedly his sorrow was considerable. Isber, the poor boy, had wanted to grow up to be an Ottoman gentleman; his house represented his claim to that status, to those values. But everything shifted under his feet. Ottoman gentlemen were no longer what one could strive to become. The world had made them anachronistic. As his days grew more difficult, Isber must have longed for and been haunted by what he had barely touched but not had the chance to savor. Perhaps he might have been relieved, or at least surprised, to discover that, all these decades later, at least one true Ottoman gentleman survived. "I lived the last whispers of the Ottoman Empire," Cecil Hourani told me.

When I had informed Cecil of my impending journey to Marjayoun, he was delighted that I, a grandson of Marjayounis, was planning to restore one of the town's anthology of houses — as long as it was not Isber's. Sharing Shibil's wariness of mixing real estate with family, he spoke against my decision without reserve. He had, in fact, been adamant about this, and for months had tried to find another house, any other house, anywhere in Marjayoun, for me to buy and renovate. He had even contacted relatives in the United States to gauge their interest in selling theirs, abandoned for years.

"In the meanwhile," he wrote, "my advice to you is not to get involved in the Samara house which we visited together. It will cost you a lot of money, and in the end it still won't be yours, and anyway it is not really very attractive, although it has some good features — but I think its repair would be a headache. So be patient, there are other possibilities." Trying to change my mind was one thing — many certainly agreed with him, and I could argue only out of pride — but "not really very attractive" house? This was unforgivable.

Now, at his house, the wisdom, *my wisdom,* in rebuilding my family's home came up not once. This was a good sign, I thought. Not that Cecil had changed his mind. He simply accepted my decision as a fait accompli. I was surprised. I had expected a reprimand. Often brilliant, always sharp, and occasionally curmudgeonly, as befits a British-born man in his nineties, Cecil was not only an Ottoman gentleman, he was Marjayoun.

After a few moments of hesitation at the start of our encounter — in which we both seemed to have decided not to bring up anything ticklish — Cecil brought me outside to the garden, to share *tabouli,* garnished with lettuce, and an eggplant dish known as *moutabal,* which we had picked up together. A green-tinted jug of homemade arak, with two small glasses, awaited on a stone table, along with olives, oil-drenched balls of *labneh,* and bread.

Near our bench were red aloe cactuses. Jasmine, white blossoms skipping along the vine, seemed to prop up the stone wall. From his balcony, the mountain and the towns of the Arqoub unfolded beyond us. It was a lovely setting, basking in a sun that eventually let me take off my scarf.

It is the shades of green that make the valley so lovely, he told me. I nodded in appreciation. In a few words, he had captured what I had long tried to understand about this landscape. The hues contrast with the cream colors of the stone, he said, and the feminine cadence of the topography. "I call this a lost valley. In a good way," Cecil said. "It's undiscovered, and I hope it will survive."

Slight and frail, with a prominent nose and narrow eyes that seemed calculating, Cecil had lived a picaresque life in a family whose exploits, pedigree, and accomplishments refused comparison. He was born in Manchester, England, to Fadlo Hourani, a prominent merchant who emigrated there in 1891 but who never lost ties with his birthplace, eventually helping found perhaps its most prestigious institution, the Marjayoun National College. His mother, Soumaya Rassi, was from a prominent family in the neighboring town of Ibl al-Saqi. Their children were raised in two worlds, which Cecil captured in a memoir written nearly twenty-five years before.

Thus to my earliest memories in Manchester, where my three sisters and two brothers were born before me, there were two faces: the one Near Eastern, Lebanese, full of poetry, politics and business, the other partly Scottish Presbyterian, full of Sunday churchgoing and Sunday school, partly English through an English nanny and a succession of English and Irish cooks and maids. Nothing epitomized this dichotomy more than the diet on which we were raised: on Saturdays, when my father lunched at home with his Lebanese and Syrian fellow businessmen and clients from abroad, we ate the food of the Lebanese villages — kibbe, and the traditional dish of Saturday, mujaddara, or Esau's pottage: on Sundays there was an English roast, followed by an apple pie or a milk pudding.

His British birth and distinctly British education at Oxford aside, Cecil wholeheartedly adopted the eclectic preferences of his ancestors. Soon after World War II broke out, he ventured east. Still fixed in his mind is the unbridled chaos on the quays of Alexandria as Egyptian porters fought to unload passengers' goods. In Marjayoun, where he stopped next, he met his father's sisters. "*Taqbourni*" (May you bury me), they said to him over and over. It was, Cecil said, one of the few Arabic words he understood at the time. Next he traveled to wartime Cairo, serving in the British army, then as a diplomat with an office the newly formed Arab League set up in Washington, in the decisive years around Israel's creation. After a stint as a professor, he became an adviser of fortune — to Tunisia's president, then to Prince Hassan of Jordan. Well past retirement age, he undertook, of all things, a campaign to help restore the Albanian monarchy, claimed by King Leka I, in a referendum in 1997. (He was unsuccessful, and the restoration was rejected.)

By the time I met him, his official career was coming to an end. These days he spent weeks at a time, sometimes longer, in Marjayoun, at his ancestral home, with its blue shutters and antique doors and magnificent garden.

Cecil was what might be called a Marjayouni nationalist, someone who devoted great attention and energy to a town he adored but found difficult to fathom. He visited far more than most of the town's expatriates. With Hikmat, he was elected to the town council, and in the aftermath of the war in 2006, he searched for ways to revive Marjayoun, from starting a campaign of solidarity with neighboring villages more damaged in the war to the renewal of a movie theater, the founding of a farmers' market on Fridays, and the reopening in the Saha of the Akkawi Restaurant. The war, he declared in a letter to the less enthusiastic, was an opportunity.

He wrote: "I suggest we take these events as a challenge to our will: to assert our determination to demonstrate that Jedeidet is not 'withering away,' but very much alive, and able to surmount our problems. It is now our opportunity to re-assert Jedeidet's historical role as the center of our area."

He seemed to speak with authority about anything, and he had done everything.

"You lived a Levantine life," I told him, "that you couldn't live today." Cecil was old and his movements were slow, but I could see him slightly turn his head, his eyes flashing a hint of acknowledgment and agreement. He did not agree all that often; at his age, he said, he could say what he wanted. But I could tell the comment struck him. His was a Levantine passage, and the Levant, his Levant, no longer existed. "This is a society which has now vanished," he said. "Those whispers of the Ottoman Empire have finally expired everywhere."

The range of Cecil's career and interests was Levantine; his family seems almost a portrait of its qualities. And the Houranis, scattered around the globe, shared a culture that transcended borders and simple, often artificial, notions of a nationality. They were not an anomaly: Families of earlier days often had branches in Iraq, Lebanon, Egypt, Syria, and Palestine, when they were not yet countries but shared a borderless, sophisticated culture.

I asked Cecil if he missed those days. He shrugged, not one to reminisce. They lingered. All the might of Europe's imperial powers could not completely extinguish the connections, routes, and trajectories of the Ottoman era. But generations had passed. Cecil told me he still had a few friends who might be called Levantine, dispersed as they were.

"But these families are fast disappearing," he said.

"I try not to regret the past," he added, stubbornly. "The past is past."

We finished lunch and took inside what was left of the food and arak. Next Cecil wanted to show me more of the garden, itself an expression of older values. With a barely perceptible enthusiasm, he grabbed his cane and we began slowly walking. As I looked around, I marveled at the garden's size, sprawling past what the eye could see, step after step of terraced land bringing an even greater sense of distance. There were figs, pomegranates, and olives, five or six of them planted by him, thirty-five planted by his grandmother from Ibl al-Saqi, Um Fadlo, who died in 1926 at the age of ninety-nine. The trees themselves represented more than a century.

Cecil's gardener, Ali, picked the olives and kept half the harvest. "He's quite happy," he said. Orange trees abounded, of so many varieties I lost track, their fruit ripening in the winter — Valencias, navels, and mughrabis, resembling blood oranges. Lemons, too. We walked on, and he navigated the buckling stones with his cane, dressed in his tweed jacket, gray pants, and black leather shoes. "Imagine," he said. "Someone who has grown up in England to have a garden with lemons." He laughed, a low chuckle.

Along the wall stood enormous cactuses, the branches so old that their smooth green textures had been replaced by the hard, desiccated bark of ancient trees. A Palestinian friend had told me that the same sort of cactus served as reminders of some of the hundreds of Palestinian villages that Israel had destroyed in the 1948 war and after. Those villages had been bulldozed, their names erased from the map, and their memories consigned to family histories. But the cactus roots ran deep, and year after year, generations after the exile of their caretakers, they would reappear, summoning long-lost fields, gardens, and homes. They were remnants of dispossession, testimonies to survival, just as Cecil's trees and flowers were statements of grace and cultivation in the midst of turmoil.

When we returned to his house, Cecil brought out a piece of *darabzin*, or iron railing, the kind that girds balconies and windows. Its black paint was worn away, and a few joints had pried apart, bending the rods like a hanger. It was still intact, though, and given its age, it remained in remarkable shape. In an understated way, almost as an afterthought, Cecil mentioned that Kalim Qurban had given it to him as a gift in the late 1970s. I shook my head, trying to recall the name. Finally, smiling, I nodded. Decades ago, Kalim Qurban had bought the house from Isber's brother Rashid, whose home adjoined his, sharing a wall. They had been neighbors.

The *darabzin* had come from Rashid's house, and its design was identical to the balustrade that still hung precariously off the balconies at Isber's house. Cecil wanted me to have it.

"This is my present to your forefathers," he said.

There was little emotion in February when Nabeeh and his sister Nabiha left. Bahija put in his pocket gold coins she had taken down from the attic, adding to the money stashed in a sack, tied tightly with string, that Isber had already given him. Isber spoke to the driver of the horse and buggy, urging him to be careful.

Riding in a horse and buggy as a boy, Nabeeh had thought he stood atop the world. Now, as the horse plodded down the dirt road, brother and sister looked back at the house's cream-colored stone and green shutters. Nabeeh would not see the house again for more than ten years. For Nabiha, it would be forty years.

Three days later, they were in Beirut, and on February 18, they left the port in a ship named the Lotus *for Marseilles. There, they trekked the same way other Lebanese had, riding the train to Le Havre, their port to America. They boarded another ship, the 580-foot French steamer and veteran of World War I named* La Savoie, *for New York. Nabeeh and Nabiha were numbers 29 and 30 on the ship's manifest.*

On the deck, many of the passengers seasick, Nabiha played with her first cousin Edna Abla, the only two girls to make the trip. Nabeeh sat with the men, twenty-two of them in all, bound for America. In three weeks, they arrived. On March 24, 1920, a date Nabeeh would recall effortlessly from that moment on, he disembarked in New York.

When the ship had passed the Statue of Liberty, her arm hoisted confidently, her torch, crown, and stola still colored a smoky copper that had yet to give way to its green patina, Nabeeh, scion of the Samara family and exile of Marjayoun, uttered words he would forever remember.

"This is God's country. This is home."

Returning after lunch to Isber's house, I was greeted by disorder as far as the eye could see, and probably farther: the indentation of the corners or closets was enough to send me to the nearest bottle of arak. The sunlight, rarely an ally, revealed unbelievable accumulations of dust which, if inhaled, seemed capable of creating serious lung damage, leading to protracted hospitalizations and legal actions involving aggrieved family members and packs of attorneys summoned by 800 numbers. I winced as I surveyed the odd assortment of things that would, one day soon, have to be packed off: window frames, a white marble sink, some metal shutters, piles of cedar planks, scraps of corrugated iron, and pieces of plastic as jagged as broken glass. A pile of sticks had mysteriously appeared, suggesting the possibility of future campfires gone out of control.

A huge rock lodged in the corner of the kitchen summed up all my frustration. Since August, Abu Jean had promised to remove it.

"Tomorrow," he told me each day.

Hassles aside, the house at certain moments held out promise. The red clay shingles now descended over part of the roof that, for eighty years, had no more than a flat concrete ceiling. George's stone wall behind the house neared its completion, rising six rows and more and showing no signs of arak-induced swaying. Inspired, he had also cleaned the stone of the house's façade and etched out the old mortar, some of it crumbling like drought-stricken soil. But then there was the tile. After word of the priest's anger at Labib Haddad — even under the threat of a vengeful God's wrath, Labib had yet to finish the church — I knew we had to find another tile layer if we were ever to start downstairs. The next day, Abu Jean, always the team player, produced Malik Nicola Jawish, taxi driver, hunter, butcher, fisherman, refugee, enthusiast of the water pipe, and *maalim* of tile, plumbing, heating, and air conditioning.

"The one who knows workmanship owns the castle," Malik told me. Or, less literally, wherever you throw him, he'll land on his feet. The proverb captured the man, and that was the best description of Malik — a *man*. It was as if his every sentence — mundane, peripheral, incisive — was rendered with an exclamation point. "I don't lie! I say four o'clock, and I'm there at four o'clock!" he thundered. "There are a lot of people in this work who have stopped, given up. Twenty-eight years ago until today, I have not stopped working for one day. Not a single day! I work for the best people. First of all, I'm honest, and my work is proper. That's the thing! That's the thing!"

After Malik finished a pattern of four interlocking tiles, he swept them with a frayed hand broom. "The most important thing is the tiles' cleanliness," he said. Cleanliness, it seemed, was part of the code of a *maalim*. Everything had to be kept in perfect condition, with smooth edges and sharp angles. Corners were always precise, lines accurate. This part of the project ran with efficiency. Abu Jean and I carried in the tile, Malik's assistant mixed and brought the cement. Malik barreled forward like a steam engine, row after row.

"You're an artist, Malik," I told him, and as he grunted in approval, I heard, in some corner of my brain, an imp of an observer silently mouthing the words *I am honest, too.*

At the end of one fairly productive day, a board propped against the entrance by Abu Jean fell and I looked inside at the progress of the room now revealed: The tile was laid, and it looked great. I stood at another entrance, for another perspective.

As I stared at the Cave, I had the same feeling of descending toward the Litani River, overwhelmed by the valley's view. I still saw sand everywhere; cracked tiles were stacked in two piles in the corner. Broken pieces lay here and there, their shapes indeterminate, since we had electricity but no lights. But the house — at least a very small part of it — finally approached the unique place that it would become. I felt a sense of triumph, albeit small.

There was one concern: It had taken us four months to finish one room.

Soon after, I heard from a friend that Labib had found out Malik was finishing the tile. He was angry, apparently inflamed with the kind of resentment that comes from betrayal. "He fucks ants and milks them," Labib had said of me.

These charges had not been leveled at me before.

11

KHAIRALLA'S OUD

On a foggy night in December, I sat with a friend at a bar in Ibl al-Saqi, snacking on peanuts, slices of red apple, and pumpkin seeds. We were hovering near an *ujaa* that was burning *jift*, crushed olive pits, though the flames felt Sisyphean in their struggle with the cold outside. In the corner, the television went on, as it always seemed to, blaring agitprop and trash talk. I heard only snippets, words that said impasse — mentioned were the United States, Syria, Iran, and Israel, crises and elections. Rumors somersaulted across the country. There was talk of protesters preparing to fire on police and soldiers; of a stalemate lasting months, maybe longer; of the possibility of civil strife. Leaders in Beirut speculated that a civil war had already begun. That day, a parliamentary session called to choose a president was delayed for the seventh time. The office had been un-filled for months, an absence paralyzing the state. The French foreign minister assured everyone that a deal was within reach. But all knew that any agreement remained far away.

Soon the word *ightiyal* joined the drumbeat. Assassination, it meant. Dead was a general, François al-Hajj, killed by a remote-controlled car bomb that ripped through a busy street overlooking Beirut. Broken glass sprinkled across the pavement caught the morning sun. To the culprits, shrouded in anonymity, killing was part of the country's political calculus, the cheapest way to reach the audience. So that night in December, as on so many others, the voices thundered on and on.

For so long, Lebanon had wrestled with the rudimentary questions of identity: whether its inhabitants were Arabs first or Lebanese above all, whether they belonged to East or West, whether they were bound to a destiny that stretched far beyond its borders — the Muslim world, for instance — or were part of a legacy as particular as the history of ancient Phoenicia. A generation or so ago, there was left and right, atheist and devout, radical and reformist, unreconstructed Maoist and millenarian Salafist, sometimes sharing a table along Hamra Street in Beirut. All this combustibility ended in civil war in 1975, though for a brief moment that preceded it, Lebanon, always terribly small, had become a grander stage for ideas and struggles. For years, it was sanctuary to Arab dissidents, Palestinian exiles, and Lebanese of all sects and tendencies, who in turn delivered a cadence to Arab politics in an environment that, while ridden with strife, oppression, and poverty, was relatively free and unencumbered.

These days, grander struggles had given way to narrow ones, and Lebanon had become much smaller than it ever was. Now it stood mired in the confrontation that began after the 2006 war. The conflict never really changed, so much about it static, except the sense that it always seemed to be hurtling toward some precipice. Tired clichés reappeared again and again, month after month, in a discourse too inflexible for the inclusion of possibilities not already tiresomely rehearsed. And besides the pettiness, it all seemed a sad confirmation of Lebanon's loss of intellectual boldness. Only the most devout were imbued with any sense of real politics, if politics is to mean changing societies and the political orders they have given rise to. Lost these days was any notion of collective action. Bonding around principles or ideals seemed a kind of romantic memory, the stuff of archived posters with their dated haircuts, images of Kalashnikovs, and clarion anthems that seemed nostalgic yet relevant. Even as tectonic shifts seemed to rumble in the Arab world around it, an imagined spring after the longest of winters, Lebanon had little notion of belonging to heal or inspire, save religion and its burden of the sacred. Dialogue became weakness, and empathy the stuff of craven apologists. There was no debate in fundamentalisms, save the squabble over trivia or abstractions until the guns moved in. All too often, as on this night in December, we were left with spectacle.

Tension gripped a little tighter as the year ended, and though swaths of the country were being rebuilt after the losses of the 2006 war, the hammers, bulldozers, saws, and drills of the construction in progress seemed almost drowned out by the cynical, dramatically extolled invective of sectarian leaders. These old warhorses knew their parasitic relevance depended on conflict, which of course concealed the real questions. On television, a Christian-owned station introduced broadcasts with the ominous phrase "the silence before the storm." That sense of premonition, of things still reluctant to reveal themselves, was familiar now and then.

In 1911, nine years before Nabeeh and Nabiha Samara left Lebanon, Ayyash Shadid died, leaving his wife, Shawaqa, and their children in a simple stone house with a mud roof near the Serail. The oldest child, Miqbal, had departed Marjayoun some years before all the others. He fled, like so many, because his mother could not bear to see him die for the Ottomans on a faraway battlefield, be it the arid climes of the Sahara or the plains of Thrace, in the conflicts of a fading empire. Miqbal escaped the draft, but his mother lived in fear that her second-oldest son, Abdullah (who would become my grandfather and was then nearing eighteen), would not. Shawaqa insisted that Abdullah act quickly and join his brother in America.

Less than a decade after Miqbal's departure, Abdullah followed, taking his sister Adeeba with him. Carried by donkey, they fumbled across deep valleys bisected by the Litani River, then trekked for days over the mountains before reaching Beirut, where they had barely enough money to buy transit aboard a ship. In France, they found passage on the Latso, which arrived on October 1, 1911, in Boston, where one charitable group declared that "next to the Chinese, who can never in any real sense be American, they, the Syrians, are the most foreign of all foreigners."

In time, Abdullah would change his name to Albert, and his signature on the legal document was shaky and unsure. His penmanship was like a second-grader's, the scrawl in a language that he managed to master as an adult, more out of pride than necessity.

After Abdullah got settled in America, both he and Miqbal sent money home to Shawaqa, but during the long years of World War I the mail was interrupted. When at last it resumed, a letter arrived. "To the wife of Ayyash Shadid," it read, in handwriting Shawaqa recognized, all these years later, as Miqbal's. "If she's not alive, then one of her children. If her children are not living, then one of their relatives." In the package to which the letter was attached, Miqbal had sent clothes and money, neither of which Shawaqa and her children had much of. Make sure to buy candy for the youngest of the children, ten-year-old Nabeeha, he instructed.

At the time, Bedouins, defying French authority, were raiding villages in the Hula Valley and along the Palestinian frontier. They also made repeated raids on Marjayoun, torching the Serail, charring its cream stone, papers fluttering out its windows like confetti at a parade. The raiders eventually reached the Shadids' road. Ayyash's house was burned. Gone were the clothes Miqbal had sent; the remains were strewn along the road or piled in charred, coarse clumps in the yard. The brigands shattered jars of food and oil on the rocks, leaving shards of pottery tucked in the grass and encrusted in dirt.

In the wake of the upheaval, the family — which consisted of Shawaqa, her daughters Najiba and Nabeeha, and her son Hana — decided to leave, joining the others in America. Miqbal sent them $1,000, and they became part of one of the largest groups to leave Marjayoun. For three days, they walked to Beirut. Leading them was the Reverend Shukrallah Shadid.

In Beirut, they took the Red Star Line, the children qualifying for half fare in steerage, where one of them would recall that they traveled "like cattle." Twelve days later, they were in Marseilles, where they stayed a few weeks, trying to find a ship to take them to America. Booking passage in the bottom deck of the LeBlanc, they soon headed for New York City.

On October 5, 1920, his will in hand, Shukrallah Shadid arrived with his family and the others in Oklahoma City, their train pulling into the Santa Fe station. "America is a new country, a free country," the priest told a sixteen-year-old nephew, who would remember his words long after. "You can be a force for either good or evil, as you choose."

"They say Lebanon is the Switzerland of the Middle East," Dr. Khairalla's wife, Ivanka, told me later when I stopped in at their house. She said it with a hint of sarcasm, although her words were more tender than bitter. "And Cecil says Marjayoun is the queen of Lebanon." All three of us laughed, though Dr. Khairalla was perhaps a bit embarrassed at the words about his friend. He turned to the window for a glance at the winter light, unusually sunny. Ivanka was decorating their Christmas tree, a splash of color in a house shaded in browns.

Dr. Khairalla was gaunter and paler than when I had seen him last. There was a fatigue about him, too. I did not know him before he had cancer, but friends had told me he was far more outgoing and exuberant, generous with a laugh. I might not have recognized that man. To me, he always seemed meditative, drawing on a weary dignity.

I spent more and more time with him and had promised to come to his basement workshop, where he was building a violin, the instrument his wife had played as a young woman in Bulgaria. After coffee on this December day, we descended the stairs to his shop. On the wall, four lines of poetry were written in Arabic:

> *The oud's yearning voice sounds like a penitent's grief.*
> *Together, people long for it.*
> *Perched on it, a bird warbles,*
> *Longing for the branch that was once its tree.*

Shelves were crowded with dish after plastic dish filled with nails, drill bits, tape, and glue. Paper templates, models for his musical instruments, were spread across the floor, mingled with scraps of wood, kindling for heating down here. On the table, over the workbench, and against the walls, the models were strewn. The workbench, he said, pointing at it, was more than a hundred years old, and its scars guaranteed that.

Five or so ouds hung from the ceiling, along with four dangling bouzoukis. A finished oud leaned against the wall. Since I heard it as a boy, the oud has moved me. No instrument evokes such soulfulness as the pear-shaped oud. The sounds of its strings, brittle but rounded, suggest the accompaniment to a lament. By legend, the instrument dates to the start of history, when it was invented by Lamech, the sixth grandson of Adam. Lamech's only child died at just five, and the father used parts of his son's skeleton as the model to build the instrument that could convey his grief. Weeping, he played it until he went blind.

Dr. Khairalla built these instruments even better than he played them. How long did it take him to complete one oud? I kept asking. He didn't measure the time, he kept saying. It was a hobby; the harvest, not the bounty, was what mattered. "Sometimes one year, sometimes one month, sometimes two months." His answers were not vague or dismissive. If anything, they were modest and quietly precise.

The instruments were the doctor's passion. Fashioned in a medley of wood and glue and varnish, their shades were determined by the fir, mulberry, basswood, lemon, or other local timber he used. His wood of choice was fir, scraps of which he had collected over the years from the backs of old wardrobes. Each of the beloved instruments bore a nameplate in Arabic: *Made by Dr. Khairalla Mady, Marjayoun, Lebanon.*

Dr. Khairalla showed me a two-stringed bouzouki beautiful even by his standards. Its lower part was chiseled from a single piece of wood. "This takes time," he said, pointing out what he considered his masterpiece, an oud whose rounded bottom was crafted entirely of matchsticks, thousands arrayed in the most stringent of longitudes. He had worked on it at night, over two or three months, "during the first war in Iraq."

Time, for the doctor, was often refracted through traumas: He became a doctor before the civil war, returning to Marjayoun as it raged. He began at the Marjayoun Hospital during the Israeli occupation, and he left it after the occupation ended. He had a plan, but thinking everything out beforehand doesn't work with wood, where craft is guided by intuition. Dr. Khairalla never apprenticed as a maker of instruments. He relied on inspiration.

"What you see now is what I imagine," he said, not the sort to elaborate much.

When a woman asked him to sell her one of his bouzoukis, so that she could give it as a gift to her son in the Persian Gulf, she offered a thousand dollars.

"If you pay me ten thousand, I will not do it," he told her.

As we walked upstairs, I saw more instruments crowded on a bench in the stairwell — nine in all, I think. Five were ouds, and he pulled them gingerly forward to reveal a peculiar bouzouki made of three gourds. Its wild, unkempt look suggested the adventurous. He stared at it as if he had only just noticed it. "Interesting," he said. "During the summer, I'll make something here to collect all of these."

He cradled that feral bouzouki, tuning it for a few minutes, then strummed a song of passion. It belonged to Fairuz, a Lebanese diva, and he sung these words:

> *O Laure, your love seared the heart*
> *After I had given you my love and affection.*
> *Remember the fields of youth*
> *And the pledge we made there to stay loyal . . .*

He whistled the tune, still strumming, then set the bouzouki down.

"Next summer," he said again, "I'll find a place for it," and the tone of his whistling became almost plaintive, like a goodbye.

The Shadids who left on the Red Star Line were Isber's neighbors, their house a five-minute walk past the Gholmias and other Samaras. As children, Nabeeha and Najiba had played with Isber's daughter Raeefa, and Isber was reluctantly, grudgingly assured that they were making the same journey that his two eldest children already had. Perhaps Raeefa could go as well, Isber thought.

Just twelve, Raeefa was too young to have the same conversation with her father that her brother Nabeeh had. This time Isber would decide, and as he sat with Bahija, whispering, exchanging as few words as possible, they pondered whether or not she should leave. Raeefa, whose name means kind and compassionate, was the oldest still in the house, but vulnerable. Destined to be small as an adult, she was nevertheless tall for her years, making her look older. This worried Isber; she was no longer a child who might pass untroubled or untouched in the countryside. But Isber could not send Raeefa off by herself, whatever happened in the town. She was just too young. He would try to cheat death again, he thought, as he had in the Houran, as he had with the Ottoman draft, and as he had with World War I. He would wait.

His period of indecision proved brief. One day, his sister Raheeja visited, joining him as he sat smoking on his balcony above the liwan. Five years older than Isber, Raheeja was married to a relative of his wife's, Mikhail Abla. She had come to say they were leaving to join their children in America. Their daughter Edna had traveled with Nabeeh and Nabiha. Their son Ellis was already there, settling in Oklahoma.

Could they take Raeefa with them? Isber asked.

In Marjayoun, in Isber's house, Raeefa had been raised, in the Ottoman style, to marry a gentleman of distinction and take her place in the quiet of women's lives. But this would not be her fate. She was leaving. Later, Raeefa's passage through Isber's rooms must have seemed only days or hours to the older woman whom I saw in Kodak shots.

What did Raeefa remember from that stretch of time called home, when there was, in her father's house, the feeling of no harm possible, the sense that nothing could touch her? Did she remember flashes of faces, the embroidered dresses stitched by her mother, the shine of everything Bahija could buff or polish, the balcony where her father ruminated through nights of fires blazing on the mountains? Eddies of smoke from his cigarettes curled and drifted by day into the clear air of the valley where the Litani River released its currents.

From her window on the morning she was to go off to America, Raeefa might have looked out, noticed the fine buggy, almost certainly shiny, that her father had hired to carry her away from him forever. But she had *no idea she would ride away that day. She was just twelve. The buggy was the last gift her father could offer.*

Whatever Isber said to his daughter on the day of her departure, I will wager that he said it not with words, but with that shiny ride, a message she might not have heard on a day of things fast and barely taken in. Such a surprise it must have been to learn that she had slept her last night in her parents' house. Such a shift in life on a morning that started out safe.

Whatever it was that Bahija said to her daughter, I would wager she said it with things stitched and embroidered, soft or warm, for a future she had no way to imagine. What mother could explain or acknowledge to herself that these things, so beautifully laundered and folded, would likely be the last of her things to carry her touch?

A few days after my visit to the basement workshop, I joined Dr. Khairalla in his kitchen. He had offered to teach me how to make *awarma,* cooked mutton mixed with the preservative of melted fat and salt. *Awarma* is a Turkish word, he told me, and the dish was prepared before winter, when neither butchers nor meat was available, especially in more remote locales like Marjayoun. Even without a refrigerator, the dish can last for months. Often it is cooked with eggs or *kishk,* a powdery cereal of cracked wheat known as *burghul* that is fermented with milk and yogurt and prepared in the fall after the wheat is harvested. It is a centuries-old tradition. "Nowadays," the doctor told me, "it's a delicacy."

Dr. Khairalla unsheathed a knife whose oak handle he had crafted. The blade came from a garage lever, a hint of ingenuity that I could tell made him proud.

The three kilograms of meat, a glistening red, were the shade of cooked beets. He dumped the sheep's fat into a pot, which Ivanka thought was the wrong size. She was assertive and ironic, and he was still charmed by her, finding humor in her needling. Not enough, though, to heed her advice. "He thinks everything he does is perfect, but there's nothing perfect. There's no perfection except God," she said, smiling.

"Now we start," he announced, and ignited the flame of the stove.

"I think there will be war in the region," he said abruptly, as the fat heated. Perched on a nearby window, his cats, Pussy and Pokhto (Bulgarian for a small bushy beard), nibbled on meat scraps. "It's very depressing. We manage to finish with one war, and we always end up starting another." There was no pity in his voice. His words were sincere, spoken by someone whose life had been shaped by war and the flight from it.

Ivanka interrupted. "It will take a little while still," she said, her voice certain. "That's my experience with war. The weather has to get a little warmer first."

"I'm not a fanatic Christian. I don't go to church. I respect all religions," he said, as if about to offer an unpleasant diagnosis. "But from what I see now, in thirty years, there won't be any Christians left here."

The television was on. The news bar scrolled with reports of more clashes between rival factions in Beirut. I was so tired of the anticipation, the dread of yet another conflict.

"It's not in our hands to stop it," Dr. Khairalla told me.

The fat had begun to melt. Dr. Khairalla took a spoon of it and put it in a wire mesh strainer placed over another pot, then pressed until the liquefied fat dripped into the pot. The smell was pungent, meaty, but staler, a bit distasteful.

He told me an Arabic maxim: "The poet can do what others can't." It was a gesture to the place that poets had in the culture of Arabs, as dissidents and polemicists, foretellers and historians, provocateurs and hagiographers. To Dr. Khairalla, it was also an insight into the arbitrary. Those with power set their own rules, he said, be it war or peace, and the powerless, with no poetic license, must make do.

"There will always be wars in Lebanon," the doctor said, looking down at the fat in the pot. "If you read history here, we've always had wars. And we may not have peace from now until eternity. I was born in war, and I grew up in war, and my children did, too."

Two days after the Israelis arrived in Marjayoun in 2006, Dr. Khairalla fled in a convoy of hundreds of vehicles, carrying as many as three thousand people. By afternoon they were crawling along a road that had been plowed for four weeks by Israeli fire. At Hasbaya, the United Nations forces ended their escort, unauthorized to go farther. Already disorganized, the convoy, in time, split up, clumps of cars moving through the Bekaa Valley toward Beirut. Confusion melted into fear. Then disaster struck.

At 10 P.M., the sky lit with a moon almost full, a flat boom resonated as the smaller convoy meandered up a slope near Kefraya, a town celebrated for its wine in more peaceful times. Pandemonium erupted, as the second car in the procession exploded in flames. Israeli forces had fired a dozen missiles at the defenseless convoy. Seven people were killed, two of them from Marjayoun — Elie Salameh, the baker, and Colett Rashed, the wife of a *mukhtar*. The next day, the Israeli military said the column was attacked because of suspicions — later acknowledged as baseless — that the cars were smuggling arms for Hezbollah fighters. Never, it said, was the convoy authorized to leave Marjayoun, although the United Nations denied this. Dr. Khairalla saw only malice in an attack more than thirty miles from the front line.

"This is war," he said again.

It was 1920 — the Year of the Twenty, it was called. No one in Marjayoun would ever have imagined what had transpired. Perhaps Isber and Bahija believed their daughter could or would forget them. Maybe that was what they hoped for her.

In his daughter's bag, Isber put the sack, tied even more tightly than Nabeeh's, filled with the gold pieces he still had from selling wheat in the Houran during the war. Bahija gave her yet more gold jewelry, knowing that she could sell it any time. "Take care of yourself," they said to her, over and over.

She could not speak or say goodbye. She cried, but could not say the words she felt to her parents. She was leaving home, the olive trees that she had played near, the fig tree that she had eaten from, the stone wall that she had climbed.

Bahija said, "God be with you." She repeated it one last time as the horse and buggy departed the house with the olive trees, carrying her daughter, the girl who would become my grandmother. "God be with you."

So many had said goodbye to Lebanon and Marjayoun by this day in 1920. Who would notice that Raeefa Samara, attempting not to show her feelings, found herself, though tall for a twelve-year-old, not up to the attempt. As the buggy she shared with her aunt and uncle pulled away from her parents, as Raeefa left the home of Isber and Bahija for perhaps the last time, the young girl proved herself unpracticed at composure. No glance at Aunt Raheeja or Uncle Mikhail was necessary to see their discomfort as they left Marjayoun for, first, Beirut, then Marseilles, and then America. But their sternness was not unpredictable. Emotional displays unsettled the world. Restraint, like proper clothing, concealed and silenced vanity. Eyes of girls were not to gaze directly, but to take in, briefly, what was expected.

Her aunt and uncle were not leaving children behind, but joining them, already in Oklahoma and doing well. And so the buggy plodded toward Qlayaa, a string of mud huts along the ridge of a hill shared with Marjayoun. As that village drifted off, the road bent three times, at its steepest elevation, then flattened as it descended toward the Litani River. Then they saw Beaufort Castle, known in Arabic as Shaqeef. A Crusader fort perched 2,100 feet above the sea, it sat improbably at the valley's summit. Behind Raeefa was Saint Elijah's, hugging a rock outcropping that looked as though a flood had swept away everything beneath it. The road tilted and turned, sliding down the valley and its sides of boulders stacked one on top of the other, until they reached the Khardali Bridge. An ancient frontier beneath Beaufort, it opened to a majestic view of the valley, as beautiful as it was lonely, hardly an inhabitant along its irrigated fields of cantaloupe, watermelon, carrots, and corn.

Saint Elijah's was still visible, like a sentinel. So was Mount Hermon and the frontier of Palestine, the Houran, and Syria beyond it. But Marjayoun had receded behind the hillside. Ahead was Beirut.

Hours passed and Raeefa was quiet, almost certainly scared as she sat next to her aunt Raheeja. The air became cooler as the horse and buggy climbed, the sun casting shadows across the valley, bordered by cliffs that were rounded by time.

The next day, as they left Jezzine, her aunt and uncle pointed out the *town's waterfall. Months until winter, it was no more than a stream. Out of respect for her elders, Raeefa smiled, but was unimpressed. From the town, the road evened out, gradually descending to the sea. The weather became warmer. Soon after Jezzine, Raeefa saw it for the first time — the blue waters of the Mediterranean Sea, tinged with green, almost like a mirage as a mist danced over it. Nothing like a river, like the Litani she knew, it seemed endless. This was beauty, like a dream. She sat staring at it, her eyes wide, as the buggy made its way through Sidon, along the coast and on to Beirut.*

12

CITADELS

NATURE STILL DICTATED much of life in Marjayoun. Even more so in winter. Every day a ramshackle car meandered through the streets with a loudspeaker mounted on its roof. "Aluminum, iron, car batteries for sale!" the doleful voice wailed, never wavering from its flat monotone, as it passed the desiccated fruit on pomegranate trees that lined the road. "Aluminum, iron for sale!" Everything around the car was gray, bathed in fog. Farther away, storms rolled in from the Mediterranean coast and on past Jabal Amil. Thunder rumbled through the valley, and then the rains would begin. Water poured through the bedraggled streets, filling every crevice, crack, and pothole.

The availability of electricity dictated everything, regulating the day — *when* the small, satellite-shaped electric heater that I called the Syrian radar functioned, *when* the three of five working bulbs dangling on a wire from the ceiling cast light, *when* the water heater scorched so aggressively that steam hissed through the shower head, *when* the mini-refrigerator kept what little was inside cool — pickles, yogurt, cheese, tomatoes, along with the fruit of the winter, oranges and clementines. (The dearth of electricity irritated everyone. "Two billion dollars in Iraq, and the Americans lit the place," Hikmat told me. "A lot more was spent here, and there's still no electricity in Lebanon?" I thought: You know it's bad when the Iraq of 2007 becomes the standard by which people measure progress.)

At night I hunched up near the *ujaa*, the cast-iron heater that had already scalded my finger, branded my wrist, and singed the sleeve of my sweater. It burned on diesel, which I left on the porch in a blue plastic jerry can. When the heater was low, I grabbed an old copy of the Lebanese daily *Al-Akhbar* and spread it on the ground. Then I gingerly filled the tank of the *ujaa* from the jerry can, which was left over from the war in 2006, when I had to carry my own gasoline for my car. For more than a year, the can had sat empty in the back, along with another blue tank and a bigger one, in a dreary olive green, a tattered piece of plastic under its red cap to prevent spilling. Now the tank had a use again. Clumsily, inevitably, when I poured fuel into the *ujaa*, some dripped onto the red Egyptian carpet I had received as a gift in Cairo for my thirtieth birthday, ten years back.

I had returned in January after a couple of weeks in the States, and Abu Jean was at the house to greet me when I trudged through the mud and made my way inside the windswept first floor I had naïvely hoped would be ready for occupancy a month before. More bluff than ire, Abu Jean began with a question.

Why had I not brought him anything from America?

"He brought himself," Toama told him, earning a smile.

Everyone was at the house. Emad Deeba, the electrician, sheathed in overalls, was working on the wiring, embedding some of it in the mortar between stones. Ramzi al-Bahri, charged with the completion of the false ceiling, fastened an aluminum carcass to the concrete roof. Toama, my neighbor and a jack of all trades — from cleaning the edges of tiles to painting the *darabzin*, or iron railing, black — was working, as usual. George Jaradi was with him. Setting hope against experience, I was inspired. Even the code of the *maalim* could not sway me.

Maalim, the Arabic word for expert or master, was perhaps uttered more than any other at the *warshe*. It was used to demonstrate pride, to convey respect, to intimidate, to end an argument, to charge more, to justify delays, and to rebuff criticism. It was both feather and club, meant to tempt and to bludgeon. "*Kiss ukhtak, wahid maalim. Maalim, maalim, maalim,*" Hikmat put it to me, perhaps a little uncharitably. "Fuck them. All I hear is *maalim, maalim, maalim.*"

"I'm a *maalim,*" any one of them would usually say to me when I questioned a decision. "I understand," I would answer, argue in vain a little more, then, once I grasped the stony facial expression, simply give up. Who was I to question a *maalim*? I suspect the look would have been no different had I asked to sleep with his wife.

"I'm a *maalim,*" Abu Jean told me as we stood staring at a pillar of crumbling concrete. The pillar's sudden obsolescence was a stroke of rare genius on the part of Abu Jean. His answer to any problem — a water leak, a crack, a crevice, a hole in the wall, roof, or floor — was invariably, "Pour cement!" And he would. On one occasion, though, he poured cement into a new bridge that buttressed the ceiling, meaning that we could remove the pillar. With it gone, the arches would assume a graceful symmetry along the entrance, an elegance that was impossible with the crumbling, honeycombed pillar in place. Abu Jean alone understood that. "The arch is like a beautiful woman," he told me, drawing his hand out as if tracing the lines of an elephant's trunk dangling from his face, "with this huge, ugly nose."

He promised to remove it. "Tomorrow," he told me.

Abu Jean was a source of fascination for me, even enthrallment. Had he simply been incompetent, I would have understood. But he only rarely was. More often, he procrastinated, offered excuses, served as his own notion of a native informant, and exhibited the most remarkable streak of passive-aggressiveness that I had encountered. My asking whether he had phoned someone who didn't show up often unleashed a philosophical diatribe. There was no point in trying to hurry them along. There was no reason to get frustrated. There was no justification for getting angry. The work would eventually get done, Abu Jean explained, whatever I did. Besides, the *maalimeen* would work when they wanted to work, whatever Abu Jean might do.

I usually stayed quiet through all this, smoldering, growing more resentful and more eager for him to leave for the day. Stay calm, I counseled myself. There was nothing menacing about Abu Jean, after all, despite his being breathtakingly stingy and unnecessarily rude to Abu Jassim and the other Syrian workers.

The other *maalimeen* at the house seemed to like Abu Jean, though their respect was sparing, then grudging.

When the rain stopped for a few days, I sat outside with George Jaradi, Toama, and Toama's wife, Thanaya. I started apologizing to George for lighting a cigarette in front of him. Always cheerful, he would have none of it. "Sit! Sit! Don't worry, Anthony." George used to smoke two packs a day, he told me. After a recent health scare, however, he quit. The same went for coffee. At first, George declared that he used to drink forty cups a day. When I looked at him in disbelief, he changed his estimate to twenty or twenty-five. Either way, he was now down to a cup or two a day. "George liked coffee with cigarettes. They go with each other, like Abu Jean and the house. Now George doesn't drink coffee, so George doesn't smoke."

It was just after 1 P.M., and Abu Jean was heading home.

"Bye," Abu Jean said curtly, trudging down the driveway, dressed for a wedding, a funeral, or a visit to Hikmat's, the occasions that prompted his Sunday best.

"It's still early, Abu Jean," Toama yelled at him.

"I'm going to eat lunch," Abu Jean replied.

"But you ate lunch yesterday!" Toama said.

That ignited Abu Jean's customary string of expletives — always disconcerting, since they came from someone my grandfather's age. He shouted, "Motherfucker, brother of a whore!"

George soon joined in. "Why so angry, Abu Jean? You look like a groom today! Where's the bride?"

Wandering in a wintry silence, Abu Jean didn't hear him.

That jovial goodwill seemed to start fraying in the ensuing days of winter. Abu Aajeh, Mr. Loudmouth, Abu Jean had called George for weeks, often the retort to anything Abu Jean heard — or didn't hear — George

say. Now Abu Jean added another putdown. He said, "His ass is jiggly."
Like so much Jell-O, it meant — or, less literally, that George was lazy.

After that, George grew less charitable. He had been sandblasting
the stone walls upstairs. Finished, he went into the yard and turned
the blower on himself, as if he was taking a shower. Clouds of dust
billowed from his pinkish red shirt, the same sweater he always wore,
and his jeans. He turned around, propped his leg up like a urinating
dog, and in another performance, scrubbed his underarms. The sand
gave way cartoonishly — from his ears and nose, from his eyebrows
and mustache, from his hair and arms. He stomped the sand off his
shoes.

From there, we went to Toama's house and drank coffee next to a
wood-burning *ujaa*. George finished his recital declaratively. "George
is going to put a big portrait of Abu Jean above the entrance of your
house, Anthony." On a tattered couch, Toama's daughter was doing
her schoolwork, afflicted by that age of adolescence so awkward that
she was reluctant to make eye contact. Thanaya was in the kitchen,
dispatching the scent of onions throughout the house. George and I
sipped yet more coffee, calling into question his abstinence, and he
declared to me that he was worried.

"There is no time to waste, Anthony," he said, turning serious. "If
you stick with Abu Jean, it's going to be two years before you finish."

I nodded in grudging acknowledgment. "I know, I know," I told
him.

"Abu Jean simply wants money," George went on, not really listening to me. "To get more money, he needs the work to last longer." He
paused. "But don't tell Abu Jean that George said that."

He pointed inside the house. Behind a wall, around the corner from
another, over a pile of sand and in front of the arch, the column I had
been trying to get removed was still standing. After all these weeks, it
was still there, taunting me, that huge, ugly nose obscuring one of the
house's most elegant features. "Fouad has told Abu Jean to take the
column out for two months. Two months! If he told George to do it, it's
gone in fifteen minutes." George seemed obsessed, fixated on the column. "The column is still standing," he said, shaking his head. "Fuck

the column! George is angry at the column, Anthony. George, himself, is angry at the column! It's still standing and it's giving you the finger." He made the gesture with his hand, his face twisted in anger.

"Ten dollars," Toama said, jumping in. That would be the cost of tearing it down.

I had hoped to move into the house over these days of winter, at least into the bottom floor, where we had focused the work. Since I had not, anything short of completion felt like a disappointment to me. The column was a sad metaphor for all that had not been done. Weeks had passed, and progress was visible only to the nearsighted. From a distance, the house still looked *maskun,* haunted, awaiting salvation.

Then, a few days later, the column was gone. No one had said a word the day before. Nothing had prompted Abu Jean's decision, at least that I knew about. The task adhered to a *maalim's* sense of time, driven by its own indecipherable inertia. One morning, wielding a sledgehammer, Abu Jean simply knocked down the column and poured cement over it — *pour cement!* — to fashion a new floor under an arch uncluttered for the first time in generations. He had buried the eyesore in the house itself. For a moment, I thought, the house was sovereign.

Abu Jean smiled when he saw me. He always smiled when I smiled. There was a sense of triumph about him, too. "We're together till the end," he told me. That gave me pause, the prospect of being with Abu Jean for eternity. He took pride in what he had done, oblivious to all the frustration. "I built this," he said, cigarette perched between his lips, jabbing his finger toward the wall of the dining room. "I built this," he said, gesturing toward the bedroom. "And I built this," he said, motioning toward the bathroom. "Who's the *maalim* here?" he asked, tapping his chest. "I am!"

"I have complete trust in you, Abu Jean," I told him.

He repeated his words: "We're together till the end."

As the day ended, I tried to appreciate how much had actually transpired since we started: not a lot, but more than I might admit. By now, each room had its own tentative character. The Cave was the first room to be finished, if finished meant that the work was done. It was still in a builder's disarray. Water dripped from the balcony overhead, and a blue-and-white bag of cement sat in the middle of the room on a patch of cemento tile. But nothing else remained from a few months before — not the plastered walls, not the cement floor, not even the lights. The rooms were rooms coming to a stumbling fruition, and I occasionally caught myself staring at the cemento, imagining myself somewhere else, as the designs of the tiles washed across old spaces in the house where walls once stood.

I noted proudly that the space through which Abu Jean often paraded had been reinterpreted. Gone were the rooms for Bahija's tenants, her nieces who lived for decades beneath her. Vanished was the garage where Albert Haddad, the Israeli informer, tinkered with old American cars on a floor smeared in blood and oil. The bathroom had disappeared, as had the kitchen. With walls gone, five rooms now counted three. The arches of the old days had returned, now the arcade of an entryway. The stone was restored, harvesting its grays, blues, and creams, its mortar the softest of browns. The house was slowly being re-created, defining itself anew.

As the work progressed, I found myself wanting more and more to live there. This house, Isber's dream of security and luxury, was really the only idea I had for a home, and on more days than not, it had seemed it would never be quite complete enough for me to move in and bring my daughter here. I sometimes considered my grandmother Raeefa, still a young girl, during her long trip to New York, suspended between places, crossing the world with acquaintances but without her parents or brothers and sisters, alone, wondering who she was. Like my grandmother, I understood questions of identity, how being torn in two often leaves something less than one.

Raeefa had followed the path of her brother and sister to Beirut, where they were met by the same eager steamship merchants and retinue of hucksters who could bribe officials for a passport and serve as middlemen for the short ride aboard a dinghy to the steamer. Bribes were necessary: half a Turkish majidi to the customs official, more to the inspector of the boat ferrying between the wharf and the steamship, and yet more to the inspector posted at the steamer's gangplank. Isber had entrusted Raeefa with enough money for all, and she managed to avoid steerage and buy second-class passage aboard the ship.

She and her aunt and uncle arrived less than a week later in Marseilles, where the opportunistic sold everything from clothes to women and a few plucky Syrians had set up hostelries for their newly arriving cousins. She stared blankly, awestruck at the electric lights, the trains, the streetcars, and the tall buildings. "Everything Turkish had vanished," one of her relatives would write. They quickly made their way by train to Paris, then Le Havre, where they booked tickets to New York, the city already alive in all their imaginations. The journey so far had been exhausting, and this boat ride was grueling. Raeefa was racked with nausea throughout the journey.

She couldn't stomach the food. Only the apples and oranges kept her healthy. The voyage to New York was no less arduous than the trip from Marjayoun. For Raeefa, they were crises filled with tears, shouts, and desperation. To each, she had her refrain: She only wished to go back home. It was a desire I could understand, the desire of those whose place had been taken away.

Images flitted across the television screen, an untethered collage of pictures that were familiar yet worrisome. Messages carrying headlines declared themselves from my cell phone, blaring Lebanon's feuding leaders. *The Syrian regime is pushing the Lebanese resistance toward civil war*, said one. *It is impossible to coexist with a totalitarian party like Hezbollah, and we don't want Lebanon to be an arena for conflicts with Israel*, declared another. More clashes erupted, in Beirut and villages that sat along the country's fault lines of sect, affiliation, and ideology. The situation reminded me of words a friend once told me in Baghdad. "We're part of a play on a stage," she said. "Life's not good, it's not bad. It's just a play."

"I can already sense it," I told Shibil. "I'm going to finish this house and the war will start. I'm going to finish this house and I'll end up never setting foot in it again."

A small pile of incense, shaped like white crystals, burned slowly on the corner of Shibil's *ujaa*. He said he had thought about calling me, to let me know he had nothing to serve me. In the end, as we shared a drink, he fetched an ancient plastic tub of pickles, flavored with thyme, peppers, and "maybe cumin," that dated to the 2006 war. "Israel was dropping pamphlets from the sky," he said, "and I was picking cucumbers." Shibil shared none of my anxiety these days. Though a Christian, he was emphatically on the side of Hezbollah; its adversaries, he insisted, were "fucking Israeli traitors." He confidently declared that the Shiite Muslim movement, along with its Christian ally, the mercurial former general named Michel Aoun, would prevail.

"Don't worry, Anthony," he said, "the opposition will stop the war."

Shibil, as usual these days, was in bad shape. He limped from the door to his chair, slowed by the leg that was hurting him, and the prescription of his self-diagnosis was failing him: He had determined that he simply needed to eat more greens. He looked as though his body was breaking down. Dressed in gray and breathing heavily, he slouched in his chair.

Physical hardship was only one of Shibil's problems. "My advice is to stay away from relatives," he said as he put a wad of lettuce in his mouth and offered me some. "The shittiest thing is your relatives." Shibil worshiped his older brother, whom he had described to me more than once as a saint. Since the lunch on the river, his resentment of his other brothers had deepened, a bitterness that he spoke of more and more. It came to a head around the time a small tremor struck Lebanon. All in all, it was a pretty insignificant affair. Sitting at the house, I had actually thought it was a truck rumbling down the road beneath me. But soon after, one of Shibil's brothers called him.

"Did anything happen to the house?" he had asked Shibil.

Neither then nor afterward did he inquire how Shibil was.

"Shibil has a temper, no?" I asked Hikmat after considering my friend and his situation for some days. Ever the host, Hikmat was eager to agree at first. Then his face changed.

"He doesn't have a temper. He's just responding to what he's getting. It's a reaction." Everyone treats Shibil like a loser, he said, like a beggar.

Whenever he asks for something, they whisper a proverb: The beggar is setting conditions. Or as Hikmat explained in his own way: "You are a beggar and you want a piece of bread. Come knock on my door, and I give you a loaf, and you say, This is it?"

It is humiliation, Hikmat said, repeated time and again, on occasion after occasion, in place after place in Marjayoun, that brings out his temper and makes him angry. I nodded, understanding. All this explained his fights with his brothers, which lasted weeks. I saw the condescension he must have sensed from the butcher, who had barred him from his store the other day after Shibil suggested that his scale was calibrated wrong. His flare-ups were a last resort to maintain his dignity. With them, he could try to preserve a little self-respect.

"He breaks my heart," Hikmat said. "This man is a lonely man. He has no one."

He may have been even lonelier these days. Hikmat told me that Shibil's oldest brother, the one he adored, had lung cancer, an advanced and inoperable kind that would not leave him much more time. I had seemed to recall Shibil mentioning that he wanted to visit a brother who was sick in Beirut, but it sounded as though the brother had a cold, and I wondered why Shibil had never told me. He was so free about everything else.

"*Al-marad al-khabeeth,*" Hikmat told me. The wicked disease.

The rumors about my clandestine existence as a secret agent continued to circulate. There was, of course, some reason for suspicion. From Kim Philby to Eli Cohen, there were, are, and always will be spies aplenty in the Middle East, with covers never too fantastic. A friend once recounted to me the story of a shaggy, unkempt beggar who plied the streets of west Beirut, as mad as he was poor. No one paid him any attention; most tried to avoid him. And then, when the Israeli army invaded in 1982, there he was, directing the traffic of armored columns. If my friend remembered it right, one of the Israeli officers actually saluted him.

The cult of the agent had become an obsession in a way, though. It was a lasting image from my first visit to Lebanon in 1991, as I ventured through a downtown that, after fifteen years of war, had become an apocalyptic swath of ruins, dust, and weeds, drearily inhabited by squatters and bored soldiers. Block after block of buildings was reduced to cavernous, skeletal remains, and in the streets stagnant water and sewage filled countless craters, disturbed only by the curious driver peering nervously down the disfigured alleys. The only splashes of color were blue advertisements soliciting business for a billboard company. *What do our boards have in common with the CIA?* they asked. *They're both all over the place.*

"Keep a low profile," Samir Abou Jawdeh advised me. The son of an intellectual who once published a newspaper in Marjayoun, Samir visited the town often.

The words he told me over dinner were discomfiting. Five people, he said, had visited him to disclose that I was an American spy and that the U.S. embassy was paying to rebuild my house. (In another version, President George W. Bush himself provided the funds.) He should be careful what he said to me, they admonished him. He grew angry as he recounted the story.

"I told them never to speak in such a manner in my house again!" he bellowed. At least, that was what he claimed. I suspected he had probed for details, inquired about my family, friends, and acquaintances, speculated about my activities, suggested other sources for my money, and wondered over just what I was plotting here, in Beirut, and elsewhere.

"Why would someone with roots in Jedeida be a spy?" Shibil asked me afterward, as we sat near his scorching *ujaa*, diesel fuel staining its metal drip tray. I was rebuilding the house, he said, and both my parents were Marjayouni. My family had history here, both the Shadids and the Samaras. "It would be someone with a different name," he insisted — Elias, Cohen, or maybe Johnson. "Shadid, what is that? Unless the CIA has become more sophisticated than we think." He thought about the idea before discounting it. "If I found out you were a spy," he said, shaking his head in disbelief, "I would cut my own throat."

Shibil's sense of it: It was the curse of Marjayoun, suffering as it did from a mixture of secrecy and curiosity, with a requisite element of the evil eye. "I think everyone wants to know what you're doing," he said. "They're sticking their noses in something that doesn't concern them. If you don't tell them, you fuck them up. They'll be so bothered, so vexed if you keep knowledge from them." He told me he sometimes parks fifty meters away from a house he is visiting and walks the rest of the way. Curiosity erupts.

"That makes them crazy," he told me.

It made me crazy. I had come to this town with the best intentions — or at least enough naiveté to claim innocence — only to be reduced, as had many others, including Shibil, to a figure of bizarre speculation and idle whispers.

Amid the din of recriminations that the crisis in Lebanon was sowing — a headline that week read, "Molotovs, Stones and Bullets in the Streets of Beirut" — there soon emerged another subject of interest to Marjayoun. It had all begun with the same article that I had written long ago for my newspaper, the *Washington Post,* that had outraged Karim and various other Shadids. I had meant it as a simple reflective piece, with a mention of the olive tree I had planted at the house all those months back. It talked about the furrows of Mount Hermon, then dusted in snow. And it offered a lament, meant tenderly, imbued with the reverence of an exile for an imagined home. "Picturesque as it is, Marjayoun is dying," I wrote then. Hardly anyone in Marjayoun had read the piece, other than Karim, his *omertà*-like loyalties piqued. I simply filed it away in my cabinet, in that old green folder, with my grandfather's naturalization papers.

Two years later, issue number seven of *Ayoun al-Marj* (*The Springs of the Field,* a play on the town's name) was published. The headline on the cover, a portrait of Marjayoun, read, "Dr. Rahal Responds to Anthony Shadid, Page 30." And there it was. Two full pages by Cecil Hourani and Nabil Rahal, a former teacher in Beirut, offering a lengthy rebuttal to my short reflection that no one had read, or would. They did, however, read Cecil and Rahal.

"Anthony Shadid's article that was recently published in the *Washington Post* deeply saddened me as it was a death notice for our beloved town of Jedeida and a prophecy for the future that, perhaps in a time not too distant, it would become an abandoned town where only the ravens caw," Rahal wrote. "Do we make Anthony Shadid happy and continue ignoring our birthplace, or do we rise together and unite to return to Jedeida its bygone glory and a hope that is now almost lost?" He went on to make a novel suggestion: The town's exiles should visit Marjayoun regularly and rebuild their homes.

"You're from Bayt Shadid?" asked a customer at the Marjayoun Bookstore, a stationery shop where I stopped each day to buy my newspaper. Someone else jumped in: "You're fixing the house that belongs to Bayt Samara?" Samir Razzouk, the shop's owner, knew me. "*He* is Anthony Shadid," he told them. And a knowing look crossed their faces, reflecting a moment of clarity that ran roughly like this: "So you're the guy who wrote the article insulting Jedeida."

And so the uproar began. I heard little else in the weeks that followed. Cecil kept a copy of the piece on the brass table in the salon of his house. Its placement seemed to say: *Just so you don't miss it.* My neighbor, who hardly ever said a word to me, expressed interest. "Did you know they're responding to you in the Marjayoun magazine?" he called out as I hurried through my door. Dr. Khairalla, perhaps out of kindness, gave me his support in typically refined, meditative fashion. "You are right," he said sadly. "Marjayoun *is* dying."

Even Abu Jean's wife, as sweet a woman as there was in Marjayoun, but illiterate, had heard about the piece. Rolling stuffed grape leaves for a generous Sunday lunch, she started shouting on her patio, "Of course, Marjayoun is dead!" She gave me the same look as if I had asked her whether it was acceptable to substitute roadkill for the minced lamb of her *kibbe.* "They don't come here from Beirut unless they want to be buried," she said. "We welcome them only so we can shove them in the cemetery!" Even Emile Tayyar, the carpenter who made the coffins of those expatriates, said so, she told me.

Abu Jean, sitting next to me, nodded dismissively and wailed a string of expletives. Then he stood up, thrusting his saggy, seventy-six-year-old ass in my face. With flair, he reached his hand around to grab it. "You should wipe your ass with their article!" he declared, drawing wild cackles from his wife.

I no longer had illusions about Marjayoun. It fell short, as had I. The town's rejection of its history, its lack of interest in itself and in a past no longer deemed important, was not just an indicator of decline but a cause: If Marjayoun's past was of no concern, how could the place register, to itself or others, as worthy of our scrutiny? The town's self-image seemed to decline with every decision. And every day left unattended fueled its shame over its diminishment. All the days of disbelief and the afternoons of sloughing off had made action difficult to consider, as the amount of energy required to catch up or change now seemed gargantuan, undoable.

Where prayers on Good Friday once rang out from the store of Ahmed Akkawi, a Sunni Muslim in Marjayoun, wreckage remained from the Israeli attack on the square more than a year ago. It seemed an important detail in the citizens' picture of themselves. Two doors, once regularly painted blue, once part of a town with no interest in fading, remained boarded up, there in the very center of town.

In the 1930s, protests rocked the town, and residents surged into the streets, headed for the Serail. "We want bread! We want wheat!" they chanted. "We want to eat! We are hungry!" Now they ritually watched the newscast at 8 P.M., waiting silently for the inevitable. If there was electricity. Interest was stirred only by meaningless drama and idle chatter. Perhaps that is the world now, but it is, sadly, Marjayoun, drifting out of circulation.

"We live here," Dr. Khairalla's wife, Ivanka, once told me, pointing to the tile floor of her kitchen. "No one lives next to us," she said, gesturing to the south. She repeated the phrase three times as she pointed in each direction. She mentioned the few other people around. "When we die, who's going to come here?" she asked. "Cecil thinks it's a paradise. How can you have a paradise without people? There are houses, but there aren't any people."

PART TWO

———

AT HOME

· 13 ·

HOMESICK

NO ROADS, NOT a single one, lead to the place where we
had gotten ourselves.

I was already spoiled by the beauty of Marjayoun's hinterland, but the scene here was still moving. I gazed at the water and
the quiet landscape as Assaad Maatouk, a chef who had returned to
Lebanon from America to find home again, tied his special knot, then
rolled a piece of soggy bread over the hook. We were fishing down the
road from an Israeli artillery piece transformed by Hezbollah into a
shrine. Cliffs formed the sheer walls of the valley, and they seemed to
touch the sky. Along the walls were caves, ridges, and ledges furrowed
by water passing across the millennia. The Litani River next to us, a
murky green, was swift.

I fumbled with my fishing pole. Then, casting, I threw the reel into
the water. Once recovered, the line snapped and I had to restring the
hook, which eventually snagged a nearby tree. Then I kicked over the
tackle box, emptying it of its contents. Standing a few feet away, Assaad
didn't seem to notice my celebration of slapstick.

Assaad was preoccupied with his own routine. He had gravitated from place to place, trying to catch fish from the flat bank, from a jumble of concrete, from a bend along rapids, from a steep hill, and from a pile of rocks. Each time, he delivered one of his many verdicts, all essentially the same: The water in the river was too low. When he looked into it, he couldn't see the silver flash that meant fish were there. There used to be more fish, and he used to catch more, he said. It's not the way it was, not the home he recalled.

Only ten minutes had passed when Assaad decided to give up.

"I hate Lebanon," Assaad had told me, minutes after we had met for the first time at his house and before offering me a cigarette and a drink. "I wish I had never come." He had traveled back to his homeland only to find disappointment, now hardened into the bitterest of contempt.

Short, with a thick mustache and heavy eyes, Assaad bore a distant resemblance to a smaller Saddam Hussein, whom he apparently imitated in lighter moments at his restaurant back in La Crosse, Wisconsin. Describing the place, he spoke of his 1942 *Casablanca* poster, his water pipes, his arabesque lamps, his hanging beads, his tapestries, and his embroidered tablecloths from Lebanon — in other words, the restaurant was a Wisconsin interpretation of a Middle East he had hoped to find again, a place now missing, if it ever existed. In Marjayoun, the table at his house reflected his day: Drum tobacco, rolling paper, a plastic container filled with white watermelon seeds, a bottle of J&B scotch, three packs of Winston Reds and a pack of Winston Lights, along with a letter from his friend Dede, with a red lipstick kiss, Rolling Stones style, planted next to her signature. In his DVD player were four clips, advertisements or stories about his restaurant. A video called "A Day in the Life of Assaad Maatouk" lay among them. It was that, literally — an actual day in Assaad's life in America. There he was, toiling in his garden planting vegetables for his restaurant or sitting on a couch, sometimes with his girlfriend, the two of them eating a dish of raw meat known as *kibbe nayye* and drinking arak. I ended up having to watch it with him not once but twice. The interplay was almost philosophical: Assaad, doing nothing, watching himself do nothing.

The other videos were, luckily, a little more eventful, especially the one-minute cooking shows we watched the day we met. In one, he was in full Arab regalia, complete with headdress and robes, looking rather kitschy.

I tried not to grin as he began to speak of the Marjayoun he remembered and the one where he was living, though just for now. "The only good memories I have of Jedeida are of the old days when I used to walk around the countryside — the pine trees, the olive trees — with my slingshot and my friends. That's about it. No other good memories." When was that? I asked. "Until I was eighteen, sixteen probably. Sixteen, yeah." He lit a cigarette, the ash casting a glow on his face.

"Now I respect my cats and dogs more than I respect the natives of Jedeida."

Assaad said that he had returned home for family. "I was expecting —" he began, then stopped, searching for words before going on: His parents had passed away, but he still had two brothers and two sisters in Lebanon.

"I needed to get back a little to a family life." He paused. "To be with them. I felt kind of homesick" — a feeling I understood. His relatives offered to help him rebuild his house, a simple stone affair of a few rooms where his grandmother had lived. "I was so excited," he told me.

From his nephew, who undertook the renovation before Assaad arrived from America, the ensuing promises came fast and furious: There was talk of cleaned stone, beautiful tile, central heating, solar panels, and painted wood, "the very best of wood."

"And he put the cheapest wood, no paint," Assaad said, beginning his litany of wrongs trespassed against him.

Now even his nephew's name brought a look of disgust; he drew out each syllable for dramatic effect. "Everything he bought was left over from other building projects. Dirty. Cheap quality." Assaad pointed to the metal guard across the windows. "See that metal?" he said. "I could rip it out with my hand."

He said the roof was like a sponge. Water cascaded through the ceiling when it rained. Stairs envisioned as stone turned out to be only plain cement. His nephew's $8,000 renovation budget ultimately skyrocketed to more than $30,000, twice what Assaad had spent to buy the entire house from his uncle. This talk hit close to home and made me so anxious I actually began to sweat. Good God, what if my house turned into a similar story.

"Because I trusted my nephew, I sent him the money, and look what I got." Assaad said he almost lost his mind when, back in Marjayoun, seeing the house for the first time, he walked under a wood portico. Overhead, his name was spelled out *in English* in garish capital letters. That was just the beginning of the bizarre architectural accents he encountered. His expectations may have been so high that even if the house had been redone the way he envisioned, he would have ended up disappointed anyway. Maybe not. Either way, he desperately wanted to beat his nephew.

"I was going to break his neck," Assaad said. "Luckily someone stopped me." He looked around the house. "Look at what they've done.

"Disgusting." He shook his head. "Disgusting."

I couldn't disagree.

Since his return, Assaad had spent his time watching old videos, drinking, and planning his next meal. All the while he festered and fumed, replaying his nephew's offenses in his head. "'He's from America,'" Assaad cried out, imagining the young man's thought processes. "'He's stupid, let's rob the son of a bitch.' *And they did.* They did a good job on me. They took my money. Thieves!"

He rolled one of his Drum cigarettes, his movements elegant and practiced. "It was all a trap," he said, "a hyenas' trap. Hyenas did a trap for me. They have no conscience, no dignity. Very, very, very nasty people."

A man obsessed, he had begun dreaming about all that had transpired, his relatives haunting him in his fitful sleep. In one, they were snakes, and he was their hapless victim.

Dr. Khairalla, who was a friend of Assaad's and one of the few people in town Assaad liked, had mentioned that Assaad had come back to Marjayoun to look for a wife, but that people here simply didn't understand him. His humor, as dry as the red wine he used to serve with venison and elk in Wisconsin, was considered odd. He smiled a lot—not an indicator of intelligence in the town. He even laughed, which added to the impression of silliness. Assaad wasn't really surprised by the reception; he was a native son, after all. "I lived most of my life away from home. They're not going to understand me. *But I understand them.*"

He didn't speak of seeking matrimony, but I got the sense that Dr. Khairalla may have been right about his dreams of romance. Assaad spoke like a man who had experienced rejection, often.

"Ninety-eight percent of Lebanese women are whores," he told me, smoking his cigarette. "They go for money and appearance. Bullshit!"

Every few minutes, he lightly spit, trying to force small pieces of tobacco off his tongue. "My cousins, they're all whores, too."

When they saw him in Marjayoun's streets, aware of his anger, his relatives, especially the women, would offer him only curt greetings. "They go like this," he said, quickly flashing his hand up. "That's how they say hello as they're passing by. Like they're catching a fly.

"I go like this," he told me, flashing the finger.

No one else was so unrelenting in his opinion of Marjayoun, though Assaad wasn't the only one to speak this way. Fahima, a woman living upstairs from Hikmat, was another outsider and often an outspoken critic. She was originally from Quneitra, across the border in the Golan Heights, and was known as a gossip, retaining every whisper, every rumor, every fragment of history. She probably knew Marjayoun as well as anyone. She had spent year after year observing, making sense of it, keeping an eye on its follies, and devising her own crazy reasons for the way things were. Fahima always struck me as rather sweet, but lonely. I sometimes thought I spent too much time with her and with Assaad. Their views reinforced my own self-pity and discouragement, but I enjoyed the solidarity when I felt frustrated.

"This is the craziness of the people of Jedeida," claimed Fatima. "I swear to God, they're crazy," she said, nodding and running through the families—the Ablas, Farhas, Jabaras, Haddads, and, to be fair, the Shadids—who in her opinion fit the description. "It's the water. The water is bad. There's something in there."

We sat at her table, and I looked at her with longing as she inhaled a cigarette. (I had quit again.) After a coughing spell, she mentioned to me that her father, Daoud, had died in 1991, three months after a doctor had told him to give up cigarettes. Quitting smoking, Fahima insisted, killed him; it would do the same to her, she assured me. I reached for her pack as Fahima watched the television, filled with images of dissent and struggle, and she shook her head.

The road beyond Marjayoun was cluttered with the memories of martyrs, among them Rafik Hariri, the former prime minister and *zaim* of the Sunnis, killed in a car bombing in 2005, and Imad Mughniyeh, a shadowy Hezbollah leader accused of planning attacks that resulted in the deaths of hundreds of Americans and Israelis. Long before Osama bin Laden, there was Mughniyeh, whose handiwork in assaults on the Marine barracks and the American embassy in Beirut in 1983 indelibly framed a cauterized image of Lebanon for much of the world. His life ended abruptly while I was in Marjayoun, when a small bomb tore through his car in Damascus with an efficiency he would have appreciated.

In a remarkable way, Hariri and Mughniyeh encapsulated the contest between the two cultures that had defined Lebanon's crisis. Hariri's vision of the country basically called for a return to its identity as a Mediterranean entrepôt, prosperous and peaceful though infused with spectacular corruption and cynical dealmaking. Mughniyeh spoke to Hezbollah's vision of an everlasting confrontation with Israel that, in the words of the group's leader, Hassan Nasrallah, made Lebanon "a land of resistance."

Neither culture offered room for compromise. "They will not take our Lebanon," Hariri's partisans declared. *Our Lebanon.* Another slogan read, "The Future of Lebanon," the words emblazoned on Hariri's portrait. What future? I wondered every time I drove by the billboard. Dead, Hariri was memory, but his myth and the romanticizing of it were pulling the country toward another civil war. Such was Lebanon's curse: It always insisted on charting its future by way of its past.

The Mughniyeh iconography was just as ubiquitous and no less incendiary. After a life lived in the shadows, his whereabouts a matter of constant speculation, he was, in death, a public icon, celebrated with Che Guevara–like billboard portraits over the words *Herald of the Decisive Victory.* Along one stretch of street the posters for his cause became more insistent and urgent: *For the Sake of Jerusalem . . . My Blood,* one read. They went on: *Our One Enemy . . . Israel* and *My Blood Is Victorious.*

On the road beyond Marjayoun, portraits of Hariri and Mughniyeh alternated, as did their visions and their legacies, future and past. The contest between those perspectives was felt everywhere.

"George is scared of the situation, really," George told me as we sat in Toama's apartment on a cold, cloudy day when little was happening at the *warshe.* George talked about politics much more than Toama or Thanaya. "Here, nothing's going to happen. We're safe here. But in Beirut? Who knows what that's going to bring. Our children there are waiting for the rifles and snipers." Thanaya was flipping through the channels, most of them dominated by news of an upcoming commemoration for Hariri in the capital. I kept refilling my coffee cup, a little compulsively. George said he wanted to lock all the ministers, lawmakers, and heads of political parties together in parliament, without exception. He would then bring a thousand kilograms of TNT, his term. "We'd blow them all up together, and we'd be relaxed." He smiled. "We'd be relaxed."

George looked at me scribbling in my notebook. "What are you writing?" he shouted, falling back into his chair. "George didn't say anything!"

Soon after I had coffee with Fahima, Assaad and I had planned to go fishing for the last time before his threatened departure. He vowed to leave often, during his fits of despondency. My mood had brightened lately, inexplicably, and the trip was, I had hoped, not going to dim my spirits. Assaad's words could be contagious. I was late, which always upset Assaad. He called me, but hung up before I could answer. I called back to tell him I was on my way. "Okay, bye," he said curtly. When I arrived he seemed glum. "The same," he said when I asked him how his day was, an answer that never changed.

We drove toward the Litani, and Assaad started to recall the scenery of his childhood. His father, Tannios — a tailor with a scatological sense of humor and a store that sold shoes, scarves, thread, and needles in the Saha — was something of a legend around town, a different place than now. "There were no houses here, none," Assaad said as he gazed at the horizon, remembering old times. "We used to walk all over and back." It was a half century before, in a time rare because it was peaceful. Assaad had come of age in an interlude between conflicts, a simple time that he perhaps romanticized. "There used to be six, maybe seven cars," he said. They were all American — DeSoto, Plymouth, and Buick. As we looked down the valley toward the fortified Israeli border, marked by the tidy town of Metulla, he told me again how he and other children used to walk there to hunt, sometimes with just a slingshot.

"You know what I'm going to miss when I leave Lebanon? The olives, the figs, and the grapes." He thought about it a little more. "And the almonds."

Another moment passed. "And the wild herbs."

I felt — though I did not want to — a bit of a kindred spirit with the embittered man. It was the first time I had heard him say anything generous about either Marjayoun or Lebanon, and it reminded me of the nostalgia that is so often pronounced here, always unprompted: the longing for a peaceful but vibrant past. I wondered whether he was trying to return to a place that no longer existed. Isn't that always the case when we try to go home again?

We had taken a long road, quite scenic, and Assaad pointed to what used to be a concrete pool, probably fed by a spring, near the cemetery. I had heard about the pool, which had been there since the old days. Isber Samara had played there as a child.

"I learned how to swim there," Assaad said. "I was ten years old. No clothes back then. We were little kids. This town was very poor." No one had money for a soccer ball, he said, and only the rich had bicycles. "There were only eight bikes in the whole town," he said, "and Habib Lahoud would rent them out by the hour." He pointed to a clump of pine trees. "We used to hunt with slingshots here," he said. "There were birds everywhere. That was then." He was silent for a moment. "Things change," he said. "The birds are far away now."

Again, as we returned to town, we passed the faded pictures of more martyrs. Only the face of Haitham Subhi Dabuq, fallen on August 18, 1988, caught my eye. The boy's glasses, probably made for someone older, were too big for his face, and his thin mustache was yet to mature. His sad eyes haunted me. He was probably not all that much older than my grandmother Raeefa had been when Isber lifted her up into the buggy and she left Marjayoun on this road, though this boy's elders had a different journey in mind for their son.

Before they entered Ellis Island, her aunt Raheeja turned to Raeefa and insisted she turn over the gold and jewelry that had been her parents' parting gifts. You're a child, she told Raeefa. You'll lose it. They might take it from you at customs. You might leave it somewhere and forget it. I'll keep it with me, she assured her niece. It was more demand than request.

Raeefa's uncle Mikhail, like all immigrants, was examined at Ellis Island for any sign of illness. The authorities there told him he had trachoma, an infectious eye disease that could lead to blindness. Or, in the words of the Lebanese translator working at customs, "His eyes no good." The trachoma would prevent his entry into the United States, and his wife and niece, barely uttering a word, were deported from New York with him. Before she left, Raheeja was relieved of the gold and jewelry by an opportunistic inspector. Raised to be generous, Raeefa was penniless. Waiting for deportation, she and her aunt and uncle sat with people barred for a variety of reasons — suspected subversion, other diseases, insanity. A few spoke Arabic and offered their advice: The three could book passage on a ship to Mexico and, from there, smuggle themselves across the border. For the Ablas and their niece, it made sense. So they set out with thoughts of those who would greet them. The Ablas' children were already in America. So was Raheeja's sister Khalaya and Raeefa's brother Nabeeh. After arriving in Mexico, they managed to get to Ciudad Juárez, a frontier town and battlefield of the Mexican revolution. Ellis Abla had been waiting for them, their telegram in hand. Under the cover of darkness, at midnight, they crossed the shallow, muddy river to El Paso.

Dazed from one of the most arduous journeys of anyone they had heard about, they were in America. From El Paso, they made their way to Wilson, on the other side of the Red River, a town then in the throes of an oil boom. Raeefa's brother Nabeeh was waiting. He had taken the train from Oklahoma City to Ardmore, then a taxi to Wilson.

Hamdilla ala salaameh, *he said to her, still a stranger himself. Thank God for your safety.*

His sister's eyes welled up. She was awkward, not sure how to act. She looked down.

Allah yissalmak, ya khayee, *she said. God keep you safe, brother.*

14

A BUSH CALLED ROZANA

ABU JEAN LOOKED as he had every day. He was a man of a certain age; there was no disputing that. But I saw a sudden change: the intensity of his strength now seemed as powerful as his desire for a cigarette. His face was ruddy, but I noticed that his hair was as black as his mustache. He looked like a younger man, taut as a bowstring. In his gold-rimmed glasses he seemed to me not just some old baggage craving attention and throwing fits, but the picture of industriousness. I had to admit, grudgingly, that he had become as much a part of the house as the stone and the tile. The brown pants that he wore every day were faded and drab, but the red plaid of a shirt recently added to his wardrobe seemed to signal a new attitude. Perhaps because it was spring.

The new season was quickly gathering around Marjayoun. People had said it would be like this; they had spoken with anticipation and growing relief, though there were definitely pauses in the transition. One day would bring cold rain, the next a warm breeze that seemed out of place so early. Cows with black and white patches ambled up the road beside the house, wandering in empty fields and lazily grazing on the neighbor's land. I watched the buds gradually open. There were the white petals of the *oqhowan,* their hearts yellow. The colors of the *shaqaiq al-noaman* fell across fields. Snow remained on Mount Hermon, slowly receding and ready to surrender.

For the first time I grasped the literalness of the town's name; never had I imagined water so abundant. There were springs underneath the house, near the stone wall that ran along the house, against my neighbor's house, and toward the wall we had built with Massoud Samara, behind the house. Along the curb the water seeped. Behind the wall it trickled. Anywhere we dug, it soon bubbled, sometimes from a small passage no bigger than an ant hole.

The weather had rekindled in Abu Jean an old man's confidence in the world and in his own survival. "Whenever I say a word," he declared to me, narrowing his eyes, "everyone should say that's exactly right." He wore black leather combat boots stamped with what appeared to be some kind of date — 02 12 02 1997. The boots themselves were worn to their soft brown interior. He was working with wood used on the balcony, which he estimated to be forty or forty-five years old, but that seemed an exaggeration. As he nailed the boards into place, creating the framework to pour concrete, he cursed the wood.

"*Ya akrout.* You bastard! This is tough wood!"

Abu Jean had the ways of a villager. Everyone was *habibi* (my dear friend), *maalim* (craftsman), *khayee* (my brother), or *ainee* (my precious friend). "Here, at this angle, *ainee,*" he told his apprentice. The nails Abu Jean pounded looked like the wood — worn, crooked, too mangled to be useful. "The wood is tough!" he shouted again. "The nails won't go down." But they finally did. After he had one part of the frame done, he gave it a tug. As for me, I was an interloper, someone in the way of his work. It reminded me of something a colleague once told me: I was as welcome as a prostitute staying for breakfast. With each step, I seemed to irritate Abu Jean more. "Watch your head," he told me authoritatively.

What he meant was: Don't you have something else to do?

Each time I saw Abu Jean work in this fashion, my respect, for a moment at least, and however grudging, grew. With a cigarette hanging from his mouth, that long ash not daring to fall, he straddled two wood beams, twenty feet off the ground, that sagged with each step he took. Just below him was the rickety iron ladder he had ascended. He seemed at home trotting back and forth. With a hint of admiration, Toama looked at him. "Abu Jean is bent on falling today," he said, but Abu Jean knew where and how to place everything. There was precision to his work, and he labored with confidence. With an iron vise, thoroughly rusted and stained with concrete, he fastened more wood, then hammered it into place.

"Beautiful, Abu Jean, but what happens if we have another earthquake?" I asked him, joking. "If there's an earthquake," he replied, "I'll hold up the wall like this." Like an *abaday,* that tough guy of Lebanese lore, he hoisted his hands above his head, posing as Atlas. As he did, part of me thought he might try.

The end of the day arrived, and Abu Jean finished close to 4 P.M. He knocked out a board over which he had poured cement. The concrete held. Remarkably, it appeared to be almost straight. He looked at me and beamed. "Do you want to learn?" he asked me. "*Yalla,* learn."

On good days we drove through Qlayaa, and he told me again and again, with the exuberance of novelty, how residents there built their houses. "They put their hands behind a cow's ass, let the shit fall in them, and then plastered the walls with it! That was Qlayaa," he insisted. "Then the Israelis came, and with them the hashish, opium, and they made money. Now this is Qlayaa. *Ya akhu al-sharmouta!* That brother of a whore." On bad days we argued and shouted until he stormed off. One day my temper got the best of me, and I threw my car keys on the ground. I was completely exasperated. Abu Jean had barraged me, but even as I did it, I knew I was wrong. In the weeks that followed, he mentioned the incident almost every day. He was older than me, after all. And what he meant was that I was too young either to raise my voice to him or to lose my temper. Whatever happened, he was an elder, and with age came respect.

He followed this with what felt like a plea. I was like his son, Jean, to him.

"There's no difference," he told me.

I knew he meant it. From that day on, my relationship with Abu Jean changed. I learned never to ask him to do more than one thing at a time. I learned to let him deliver his expositions, from start to finish, without interruption. Most important, I learned the elusive meaning of *boukra,* Arabic for tomorrow.

"We'll break this tomorrow," he declared about a patch of shoddy cement on the front porch. "We'll pour the concrete tomorrow, too, and there won't be anything dirty afterward. Tomorrow? You'll be able to walk barefoot from here," he pointed to the patio, strewn with trash, "into there," he said, gesturing toward the house.

"We'll do that and I'll hand you the keys!" he shouted. "Does that please you?"

I nodded my head. So much to be done tomorrow, I thought.

"All I want to do is please you. When your head is relaxed, my head is relaxed," Abu Jean said. Smiling, he went on. "What did I tell you about my son? You and he are one."

Tomorrow never came, though a revelation did. Soon after, Abu Jean helped me plant seeds of a variety of curly cucumber known as *miqta.* Along three furrows, he did it with precision, stooped over, dusting each seed with a bit of dirt. Finished, he stood up, stretching his back, with a sense of triumph.

"Tomorrow, the vines will grow all over the place!"

As he said it, I understood perfectly. Finally. Tomorrow was only the future, and what was to come was always ambiguous here.

So began a beautiful friendship.

"Is this right, Abu Jean?" I asked about any one of the house's hundreds of details.

"Right? Sixty rights, *habibi!*"

Then, "Put your hand in my arm," he said, and off we went. Our arms locked, we sauntered ahead. He showed me the balcony, where my great-grandfather had surveyed the beauty of the landscape and where he had watched for marauders. It was finished. Then he moved on to the wall he had completed between the bedroom and bathroom, along with the entrance, its sides tilting toward perfection. As we walked, Abu Jean started singing the song that precedes a wedding:

"We brought the groom and we've come, mother of the bride, here we've come!" It was Abu Jean at his most endearing. He really wasn't the foreman anymore. He was *ammee,* my uncle. He never really cared about the house, but he unquestionably liked the friendship. He found my Arabic endearing; it somehow made me clever. Each time I took notes about something, he was convinced I was penning some beautiful line of poetry, and he urged me to read it aloud. He showed up every day to make me happy. Abu Jean cared about me.

Standing with George, Abu Jean offered me a cigarette. I took it. I winked at him and Abu Jean smiled. I nodded my head to the left, in thanks, and Abu Jean smiled again.

"You guys are *maghroumeen,*" George told us. "Lovers. You two are Romeo and Juliet, right out of the play. You've staked out a place in his heart, Abu Jean."

Abu Jean smiled, not hearing a word.

Yet while we had reconciled, the muted, muttered resentment the rest of the *warshe* had directed at Abu Jean gave way to perfectly enunciated disdain. Trouble had brewed in the winter, it erupted in the spring, and these days no one had a kind word for him. George took it upon himself to rouse the rabble, in an insurgency that began simply.

"Tomorrow, when George sandblasts the *darabzin,* he's going to stand Abu Jean right here, and he's going to sandblast him," George said as Abu Jean stood next to me.

Soon the tension forced me to become a reluctant mediator. No one talked to anyone else, and I was left to carry messages.

To Abu Jean, ever frugal, anyone who spent money, any money, whatever the occasion, was a little decadent and a tad irresponsible. (He never drove his car; he always paid on credit.) The less he spent, Abu Jean's thinking went, the better the job he did. His frugality was an asset, even if it reached excess, as when he refused to pay the Syrian workers or tried to shave 2,000 Lebanese liras — $1.33 — off the price of renting a truck. It became an obsession, too. In his eyes, the breadth of my victimization had become epic, a modern tragedy, my suffering endless.

"Everyone lies to you," Abu Jean told me, making an exception of himself, as we drove to a small factory for tile in neighboring Khiam. "But there's a God above. Twenty-five has to be twenty-five, a thousand is a thousand, and one is one."

"Their greed," he called it, never mentioning George's and Toama's names.

"Fuck the money," he said.

When I was gone, the disputes and fights paralyzed work, each party trying to collect evidence of malfeasance and rapacity in an onslaught of suspicion and *shatara,* wiliness. Nothing seemed to get done. Soon, George and Abu Jean weren't talking. Toama tried to avoid Abu Jean, as did the electrician, plumber, painter, and *maalim* of ceilings. Abu Jean yelled every few minutes, ever combustible, if uncomprehending. I was never sure whether he was feeling the pressure (doubtful) or playing the part of a foreman (more likely). These days, only Malik, the fearsome tiler, as stocky as a drill sergeant, engaged Abu Jean, and even then, only on the sidelines.

"Did you bring me Viagra?" Abu Jean asked when I returned from Beirut.

"Abu Jean! You don't need Viagra!" Malik shouted. "Just eat peanuts. That's all you need, Abu Jean. Just eat handfuls of peanuts! Every day!"

Abu Jean shook his head and said, "The tire is out of air."

Malik had become a force, working with a relentlessness not always shared by the others. His Elmer Fudd hat, buckled under his chin, was a whimsical touch, though he still intimidated me. I rehearsed several times how I would ask him to change the pattern of the tile in the salon downstairs. In the end, I figured a tough request would be best. *This is how I want it. It's my house, after all.* But when the moment arrived to ask him, I couldn't bring myself to act tough.

Maalim Malik, I said sheepishly, I'm going to ask you something that's going to upset you. It's the only request I'll make, I'm sorry to even ask, but if at all possible, can I add a few tiles here, here, here, and here? I braced myself for the answer, my jaw clenched. Do you have enough of those tiles? he asked me. I nodded hesitantly, as if I was a child being scolded by my mother. No problem, he said. He nodded. I exhaled ever so slightly, relieved.

His gruffness aside, though, Malik was, to me, the man at the *warshe* who truly deserved to be addressed as a *maalim,* and I always did. Malik demanded it. His pride was fierce, as was his dignity. "If I say Tuesday, it's going to be Tuesday. If I say Wednesday, it's Wednesday." And it was.

Like Dr. Khairalla, Malik appreciated craftsmanship as an expression of grace, and as a *maalim* in its truest sense, he had a respect for the material that went into the labor. I asked him how long the tiles would last after we set them down.

"A thousand years," he said. "If nothing happens to it? Generations. It will last generations. It's the same as this." He pointed to the stone on the wall behind us. "The stone, the tile, it's all the same."

Much remained to be done in the most traditional presentation of Bayt Samara: the upstairs with its *liwan,* triple arches, and marble floors. But with Malik working as he did, the first half of the house had already come together.

As Malik worked upstairs, *maalim* Fadi Ghabar cleaned the tile below with a bulky machine. *Columbus,* the machine's label read. With it, passing back and forth in a circular motion, he drove away the dirt of generations, throwing up a mist of dust in the Cave and elsewhere. Again I discovered the brilliance of the tile's patterns and colors. In each incarnation, the tile unfurled itself — first when Malik laid it, then when Fadi cleaned it — underneath the two stone arches that served as a portal to the house's future.

Whenever I wandered through these days, I felt solitude, the same feeling as when I first entered the house so many months before, as a stranger. It was quiet, perhaps even more so lately. As the spring rains fell, I returned to the past, walking across designs that were whispers of a culture now gone. The patterns were a miscellany of lives I never knew. The survivors of so many cataclysms were in the resilient colors, or in the gullies and outcroppings running through the tiles that mapped the geography of another epoch. They were footnotes of what we had lost.

On July 19, 1925, Druze farmers in the Houran shot down a French airplane, beginning a revolt against the French, led by the iconic figure of Sultan Pasha al-Atrash. The conflict would last two years, ebbing and flowing across Syria and Lebanon, and would prove a pivotal moment for Marjayoun and the Middle East. Throughout the region, inflation had wrecked savings since World War I. For at least three years, drought had immiserated the Houran and elsewhere, devastating a breadbasket stretching to Damascus, Marjayoun, and beyond. Springs and wells dried up. Entire villages in the Houran were abandoned. Harvests had plummeted, taxes rose, and nature itself appeared conspiratorial. The winter was especially severe. In that climate, the heavy hand of French colonial rule soon inspired the kind of outraged nationalist resentment that confronts almost any foreign occupation, however benign. France's was not.

After the downing of the plane, the formidable Sultan Pasha marched through the Houran. Men rallied to this charismatic chieftain. Within weeks, he created an army of as many as ten thousand men, in a region of no more than fifty thousand people. The Druze, his religious brethren, were not the only converts. Bedouins, peasants, deserters from the Syrian Legion, and the unemployed joined the raids. One of Marjayoun's chroniclers enthusiastically estimated that Sultan Pasha had in fact raised an army of one million. "Every village he entered, he was greeted with cheers," the writer recounted. The French answered by burning and plundering sympathetic villages, providing more recruits to the nationalist cause.

Men of one sect — say, Muslim Bedouins — took orders from Druze villagers. Christians offered their backing, though the support was more pronounced among the Orthodox than the Catholics. Some Christians even took up arms and fought on the side of Sultan Pasha.

Then, in a moment, the arithmetic somehow changed. As the revolt spread, the rebels dispatched a force of three hundred men under the command of Hamza al-Darwish. From the Houran, they journeyed across an already snowcapped Mount Hermon and into Wadi al-Taym, where they entered Hasbaya in November 1925. One chronicler said they then received an invitation for lunch in Ibl al-Saqi, a mixed village of Druze and Christians down the road from Hasbaya and only a short distance from Marjayoun. On the way there, a slight provoked a war. One of the rebels told a Christian resident of Kawkaba to turn over his weapon.

"It does not suit you," the rebel was remembered saying.

A fight ensued, and a revolt that, in the chronicler's eyes at least, had been national became all too sectarian, pitting Christian against Druze. When it ended, the Druze had massacred Maronite Catholic villagers in Kawkaba, indelibly shaping relations between the communities.

Before year's end, French airplanes bombed Hasbaya, recapturing the town in December. The next year, the revolt was crushed with staggering losses. Having cast their Ottoman predecessors as brutish Oriental philistines and their insurgent foes as brigands and hopelessly warlike, feudal mountaineers, the French ended up doing irreparable damage to their own reputation. Their forces, a motley mix of Foreign Legionnaires, Moroccan spahis, and Circassian irregulars, ransacked and demolished villages. The French shelled Damascus indiscriminately with artillery and planes. Altogether, according to some accounts, 6,000 rebels were killed and more than 100,000 people were left homeless. Famine was a perpetual threat. And, as the war ground to a halt, Greater Syria remained under the occupation of 50,000 French troops.

The Levant known by Isber Samara and his generation was fading like a diminishing breeze.

"When was the best time in Jedeida?" I asked Dr. Khairalla.

"Before 1975," he answered, with barely a pause.

He laughed. "Since then it's never been good."

Dr. Khairalla had volunteered to help with the landscaping of the house. On our trips across the Litani River to the town of Jibchit and its array of nurseries, we always left Marjayoun early, around 7:30 A.M.

It was on these errands to buy plants that Dr. Khairalla and I grew to be friends. Since we had met, I was daunted by him. Simply put, he was the kind of man I wanted to be, but worried I would never become — gentle and kind, principled, ever curious. Choices didn't seem to disturb him; in the fullest of lives, the way forward was easier to discern. I felt shy around him. I was too eager to impress, too reluctant to offend. I suppose I admired him too much. Our trips were slow; far from the house's deadlines, time lost its relentlessness. Our conversations wandered across landscapes, flitting past flowers and music, Assaad and Iran, and the masters of Arab strings, from Mohammed al-Kasabji to Abboud Abdel-Aal.

The landscape was bursting with colors only hinted at weeks before, and Dr. Khairalla surveyed the bounty of his favorite season. There was the yellow of *endol*, soon to be followed by the more rugged yellow of the *wazzal*. The *shaqaiq al-noaman* were everywhere, softly shaded in reds and purples, along with the untainted whites of the *oqhowan* — none of which he could smell, having lost that sense a few years before. No longer edible, the wild spinach unfurled red blooms. We rounded a corner, navigating the river below us.

"You see here, this hill? Before the war it was so nice. Full of almond trees. By this time, everything would have been blooming, and there would have been white everywhere," he said. He looked closer. "There are some remnants of the trees, you see." At first, I didn't. Then, as I stared, they revealed themselves — trunks, no more than a foot above the ground, seemingly still charred.

"All this hill was almonds," he said. "Before the war, *yaani*."

The nurseries, some of the best in Lebanon, were set around a hill, past a disheveled mess of concrete where rural and urban were indistinguishable. As was his way, Dr. Khairalla was judicious in what he bought. His biggest purchase was probably a cotoneaster he wanted to turn into a bonsai plant. Predictably, I was less restrained. On this trip and others, I bought citrus trees — clementine, lemon, and an orange known as Abu Surra (Father of the Belly Button) — and what they call *lawziyyat* — a plum, apricot, and peach. Along with those came hibiscus and rhododendrons, roses and gardenias, wisterias, bougainvilleas, and honeysuckle. I found three kinds of jasmine. There was a pine tree and a plant from Indonesia called *ash al-nahl,* along with a slew of other plants whose names I never learned.

There are two words (and probably many more) for a garden in Arabic. A *jeneineh* is bigger, sometimes employing a gardener, with a requisite need of water, pruning, and money. A smaller version, really a plot of land, sometimes overgrown, is a *haqoura.* Despite my pretensions otherwise, I had a *haqoura* — my ambitions for it always surpassing its potential. When it came to the *haqoura,* everyone spoke with certainty, except Dr. Khairalla.

"You have to plant this week!" George told me, an urgency to his shout. There was no room for flexibility, no question of this stonemason's authority. Either I planted this week or the plants were doomed to die. It was fate. Everyone had his suggestion. Toama said I should put the olives close to the house so their fruit wouldn't fall in the street. George wanted me to leave room for tomatoes. Toama's wife, Thanaya, was worried about where she could plant parsley. Only on the *saroo* would they manage consensus.

The *saroo* is a tall, thin pine, like a cypress. I had envisioned it being planted along the stone wall, imagining a postcard of an Italian villa. Stately, I thought. Sacrilege, they replied. I soon learned that in Lebanon the *saroo* was commonly found around graveyards, and each protest I heard grew in intensity. My cousins were first to shudder at the idea of adding them. I pleaded, and they said plant one if you must, but donate the rest to the church. George said the same — you should plant them only near a cemetery — as did everyone else at the house, from the plumber to Malik. Even Abu Jean had his advice. "The old ones," he said, "say that if you cut a *saroo,* you'll bring your own death. Or you can't have kids. *Harram.* No one plants that around the house." Nevertheless, there was a *saroo* across the driveway, in my neighbor's yard. I pointed this out, a little meekly.

"He's backward for doing it," Abu Jean told me gruffly.

The garden itself, though, was never a work by jury. It was a creation of Dr. Khairalla, the only person I really listened to as I got busy.

Any time I told the men at the *warshe* that Dr. Khairalla was coming to visit, as the *haqoura* became a garden, there was the quietest of responses. Someone might straighten his back. Someone else might tuck in his shirt. Abu Jean would put out his cigarette, then lament Dr. Khairalla's cancer.

The sun was going down by the time Dr. Khairalla arrived in his gray Mercedes. Everyone greeted him as *hakim,* doctor, and there was almost a reverence in the way they said hello. Dr. Khairalla walked slowly toward us wearing a gray Irish hat, gray slacks, and black leather shoes, too formal for gardening. He always carried a saw and two shears in his hands. Abu Jean hurried to help us. Ten years older than Dr. Khairalla, he was still stronger. We started with a plum tree that bore some of the smoothest, sweetest fruit I had ever tasted, however biased I was.

"The aim is to evacuate the interior, to make it like a cup," Dr. Khairalla said. He walked around the tree, looking at each branch on its own, then from another perspective, as he might gaze at a portrait in a gallery. "This one I want," he said, pointing at a bigger branch. "This one I could take," he said. He paused and shook his head. "Maybe next year."

Dr. Khairalla never seemed to see himself as dying. His words always suggested that there was life ahead. He went about the work trying, I felt, to impart knowledge. He wanted to do it right. He wanted me to do it right. Doing it right was important, a kind of morality, a proper exercise of spirit. I asked him how he learned, and as usual he would take no credit. "It's practice," he said.

For seven or eight years he had watched gardeners prune his trees and others', and now he possessed a feel for it. With the small branches, he took action. For the bigger ones, he studied them — their angle, their placement, and how they related to the other branches. "This is going downward, so I'll take from it," he told me. He pointed at another one. "This will not bear," he said.

"Now I'm the engineer and you're the worker," he said, smiling, and offering a proverb: "Don't teach your son. Let him learn through life."

At the beginning of the wall, near an electrical pylon, there was a bush, which I had occasionally watered but largely ignored. The doctor instantly recognized it as a *rozana*. He picked a leaf and let me sample it; the smell was wonderful. I shook my head at the scent's strength, like perfume, so sharp some brewed it in tea. He knelt down, his shoes anchored in dirt, and began clipping. The movements were quick but meditative, and before long he was whistling a folk song of the same name, then he offered a verse. "Oh Rozana," it went, "she is full of beauty."

In the weeks that followed, Dr. Khairalla visited the garden often. After he had pruned the plums, he helped me with new fruit trees I had planted. Once, when he came back from a cancer treatment in Beirut, he seemed tired. His steps were a bit slower, and he was hesitant to climb hills or any stairs. He was reluctant to greet people, but surveyed the trees and bushes like an artist. Each time he arrived, he brought something new for the garden, and another idea for the coming seasons. My favorites were the crape myrtle, its colors changing with the seasons, and a passiflora, or passionflower, named by sixteenth-century Spanish missionaries in South America who thought they saw the crucifixion of Christ in its blossom. The corona, at the top of the flower, was a halo or crown of thorns, the anthers were his wounds, the styles were the nails, and the five petals and five sepals were the Apostles, minus Judas and Peter.

On another visit, he brought asparagus from his garden and two bottles of sweet strawberry wine that he had made the year before. Soon after, he gave me various cactuses, which filled a greenhouse he kept at home. He was already at work, he said, on building a planter to grow strawberries next winter. In the months ahead, he wanted to make wine from the renowned cherries grown in Shebaa.

With Dr. Khairalla and sometimes alone, I found myself spending a lot of time in the garden. Each day, I probably walked around the plants four or five times, watching roses coming out, plums and peaches appearing on trees I had planted only weeks before, flowers blooming from a clump of wild tulips I transplanted, and buds emerging on grapevines that once seemed lifeless. The petunias had taken root. So had the honeysuckle. The rhododendron bore gorgeous purple flowers, and the jasmine yielded white and pink blossoms with a scent as powerful as the *rozana*. The olive trees were full of buds, and two of the three pomegranates that Cecil had given me managed to survive, sprouting a few leaves. I learned to respect the garden, where rituals and right actions prevailed. Patience was requisite. There was redemption in silence. Seasons were restorative. A garden, I realized, heals.

An old photograph of Isber Samara, taken in the winter of 1926 in what, at this faraway juncture, is blurred and faded. It resembles an image from a dream, not the product of a camera. Staring ahead, Isber is dressed in his finest clothes. He fits the part of a zaim al-hara —*a country gentleman*—*in his dishdasha, stitched of wool from the Syrian steppe and flowing to his ankles. An abaya would have been too provincial, unsuitable for the man Isber fancied himself to be. His suit jacket is Western, a mark of his engagement with a larger world, encountered through business or the construction of his house: the men who brought his samneh, or ghee, from the Houran and harvested its wheat; the merchants from Marseilles who sold the red tile for his roof; the Italians who sold the marble for the floor of his liwan.*

In the photo, Bahija sits in a chair next to him, and the distance between husband and wife suggests the emotional reserve that marked their interactions in public and maybe in private. Bahija's clothes were as simple as one might expect; not a stitch called attention to itself. Her dress falls respectably to her ankles and covers her wrists. Her jacket, finely knitted wool, appears to be high-quality merchandise of the sort favored these days by powerful, conservative women—devoid of ornamentation or frills. The only flourish she allowed herself was the mandeel that she wears loosely over her head. Its tiny flowers are gathered together in a lively string of blossoms that thread along its trim. They are purple, like the spring flowers on the hillsides of the land where she spent her life. Her hands, which had accomplished so much work, are crossed properly in her lap, probably grateful for a moment to rest.

*Was this a farewell picture? No. This was all the family that was left af-
ter the three children had departed. Nabeeh is missing. Only the younger
Samara children surround their parents. There is Najib, the youngest
son — junior to Nabeeh by twelve years — standing behind Isber and
Bahija, looking timorous but smiling. His ill-fitting suit hangs over his
shoulders, and his white shirt is buttoned to the collar. Already his eyes
are brooding, but despite their melancholy, he will become known as one
of the town's handsomest men.*

*Nabiha is missing. Perhaps the most gifted of Isber and Bahija's chil-
dren, Ratiba stands next to Isber with her hand on her father's left shoulder. There is something protective about her gesture, as if she knows who
is on her father's mind as the photographer does his job. In some ways,
this young girl seems what is expected. Ratiba embodies her mother's
style. She wears a simple skirt and blouse and a woolen jacket, but there
are a few flourishes: A ribbon is draped over her pocket. Her long string
of pearls declares an unabashed wealth that Bahija would never have
been comfortable with. Her lipstick and eyeliner suggest not just a coming of age, but perhaps a hint of rebellion in the years ahead. If her spirit
holds. Perhaps she would be the one to break the silence of the Samara
women. Raeefa is missing.*

*Hoda, the youngest child, sits on a stool next to her father's knee. Her
face echoes Isber's stare and her mother's eyes. Her hair, tied in a ribbon
that seems somehow halfhearted, is the fairest in the family. Dressed in
a coat with a floppy white collar, she holds a bouquet of white flowers
that falls almost to her laced shoes. Together, the Samaras look into the
camera, an instrument slightly intimidating for its subjects.*

*This is a family who now knows that nothing today is promised to ap-
pear tomorrow. This is a family who has learned that the unexpected can
occur at any moment.*

*What did Isber Samara believe he had learned in the world? What
was he proud of — the house with the red-tile roof and the hallways that
traveled back to better days? The children gathered around him? Or the
other children, leading their lives in another, safer place? He had given
them lives in America, a country that Isber Samara himself might have
dreamed of conquering if his days of effort and travel had not passed.*

Isber Samara did not visit America. A year after the picture was taken, on January 29, 1928, my great-grandfather died, at the age of fifty-four. They blamed his pneumonia on the wind, especially fierce that winter in Marjayoun. It must have been, as Isber had traveled many times in gales that tossed the wheat around on his way to and from a Houran that was no longer his.

15

STUPID CAT

A T THE END of February, the USS *Cole* arrived off Lebanon's coast, staying for more than two months, in what the American government described as a show of support for a government whose legitimacy Hezbollah and its allies questioned. Whatever the ostensible occasion, its deployment unleashed a fury of speculation, pronounced with authority, a favorite pastime here. "The USS Cole Is Heading to Lebanon; the Worst Is Looming," read a newspaper headline. An act of terror, an opposition newspaper described it. In the days and weeks that followed, everyone seemed to brace for the unexpected.

"Urgent," a message on my cell phone read, as I sat with Assaad, Shibil, and a friend of his, Simon Diab. "Saudi Arabian embassy in Beirut asks its citizens, especially families, to evacuate Lebanon."

An hour later, while we ate a dinner of silver bream cooked by Assaad, another message arrived: "Kuwaiti embassy in Beirut asks its citizens to evacuate Lebanon as soon as possible." Simon, a fabulist of remarkable creativity, always claiming to be privy to the secrets of Hezbollah, Israel, the United Nations, the United States, and France, warned that Hezbollah was digging a tunnel of pharaonic proportions to Israel under Mount Hermon. It was literally the path to war, he whispered.

"Two or three months more, the situation will be fucked," Simon said.

As evidence, he told me, the Spanish U.N. troops near Marjayoun had told him they were wearing their boots to bed, ready for imminent battle.

"We're living on our nerves," he said.

As kindred spirits, Shibil and Assaad were fellow travelers and foils for each other. Together, they roamed across memories, imagined and otherwise. From then on, almost any evening I visited Assaad, Shibil was there. In time, it helped. Assaad was overwhelmed with feelings of vengeance, plotting vendettas as he did. He would hear none of my platitudes about turning the other cheek. He tired of my suggestions that we go fishing. (We never once caught anything.) He wanted something else: sympathy.

Shibil understood this. On a night soon after our dinner with Simon, Assaad returned to the saga of his swindle, rendered like a Greek tragedy, punctuated by riffs about the thieving ways of his family. Exhausted, I had nothing to add. Shibil did. "God willing, you'll fuck them, Assaad." It was uttered in the most compassionate, caring voice. It was a call to arms, pronounced tenderly and sweetly, like pillow talk. Shibil said it again, softer. "God willing, you'll fuck them."

Assaad nodded, savoring the thought.

They were perhaps too much alike, though, and before long the pettiness of the town infected even their friendship. So did Assaad's foibles — namely, his unrelenting resentment of his family and pretty much everyone else. "I hope he doesn't talk about me that way," Shibil said to me. "If he talks about his cousin, that she's a whore, that means my back is not protected."

We were sitting in Shibil's house, passing the time. No one's back was protected, he told me, not his, not mine. Assaad had, in fact, complained that I was always late. I shook my head. "We're in Jedeida!" I said. "How can you be late in a town where there's nothing to do and nowhere to go?"

"I'm not in Jedeida," Shibil said in his typically disconnected fashion. "You know what I fucking mean? I lived twenty-five years here and I'm still not in Jedeida."

He had filled my glass with ice and set it on the rickety tin table in front of the sofa. He went into his bedroom and brought out a bottle of Grant's, about a fifth full. "Pour whiskey," he said in Arabic. We talked a few minutes more. "*Hutt, hutt whiskey!*"

"I don't know about Assaad," I said to him.

A few days before, I had treated Assaad, who was still postponing his departure, to lunch at Abu Charbel's, a simple place on a pretty bend of the Litani River.

He complained about the flies, informed me that the quail some guy ate at the next table was better than all the food he was served, picked at his fish with a disgusted look, and bemoaned the lack of cleanliness around him. Only the cleanliness merited my concern. As we ate, a mouse scurried by. Abu Charbel's wife chased it across the yard, and in a paroxysm of agility, stomped it to death. She picked up its tiny, mutilated carcass with the same tongs she had used to flip our trout on the grill. I turned away before I saw where she returned the tongs. Better not to know, I thought. This encounter would become the centerpiece of Assaad's conversation for days, and Shibil would hear the story several times.

"Whenever you eat with him, he talks about *khara,* crap and shit," Shibil said. "Skid marks, toe jam, fucking shit."

"Some of those conversations are disgusting," I said.

"I said to him once, Why do you mention all those things while we eat? I actually thought he was funnier than that. But he's not. He's more depressing. I have enough depression to give away for, like, a thousand years. I don't need more."

In the weeks that followed, Assaad became glummer. He had decided to keep his house, as unhappy as it made him, but he could no longer stay in Marjayoun. This time, he meant it.

Weeks to go, he told me.

Why so soon? I asked.

"I have to," he said. "I feel old, I feel tired, I feel pissed off, I feel disgusted."

Assaad spent the little time he had left in Marjayoun fantasizing about the friends who awaited him in America. Dede Mraz and Larry Dahl, he would say. He dragged out the vowel in Dahl — *Daaahl*. He said the last time they talked, Larry told him that everyone in Wisconsin missed him. "Every day, he said, someone asks, Where is Assaad, where is Assaad?"

But he was already savoring what was to come — they would cook dinner on the grill and watch the deer scamper by. "It's going to be very exciting when I get to America. Very, very exciting. There's going to be a huge party, maybe a hundred and fifty people. On the lake. West Salem, a big lake."

At this point, I had kind of had it, and I asked Assaad why he had ever left. I hoped he would give me an answer that made sense. It never did. He was a man caught between two places, one where he would always be a stranger, one where he was no longer a native. Time and change had made him a perpetual traveler, never comfortable again, like many who had lost their homes or those who had traveled across the world, always searching for them. He was like those sisters in Chekhov, dreaming of Moscow, looking for a place that might contain his great longing, never finding what had never been or what he imagined. Sometimes, it seemed to me, I saw Assaad's displacement everywhere I looked.

The evening arrived, the day before Assaad's departure. I got to his house as the sun was setting, and he seemed a little out of sorts. He was standing on his porch drinking scotch out of an oversize shot glass used for arak. "This is my second," he told me. From inside, Spanish elevator music blared from a red cassette player. The windows were open, as was the door. His cat was sitting with a small blanket in a bed Assaad had built from a milk crate. On a diet of milk, sardines, and scraps of meat, the animal had grown quite fat.

"Remembering the good times," he told me as I walked up the

cement steps. "Music from my restaurant. I get homesick." I always wondered what he meant by that word. "Midnight music," he said. He played it at night when the customers, in an alcohol-induced haze, started a party. He said the words again: "I get homesick."

We walked inside, and he told me, to my surprise, that he was getting nervous about going back to America. "I'm puzzled, really puzzled, about what I'm going to do. Last night I stayed awake thinking about it. I have a house here. I have no house in America, I have no restaurant there anymore. I wish I wouldn't have sold it. I'm going blind. This trip isn't easy for me."

I didn't know whether to feel sorry for him or get angry. All he could talk about since I had met him were his friends in America, about the good times there, parties late into the night, every weekend without fail. Now he was wondering whether it was such a good idea to return. The snow worried him — six months a year, he kept saying. He was already planning his trip back to Marjayoun when the bird hunting would be good. "I might come back in December to escape the snow. I'll see what I do. I might start a catering business. That's a possibility."

"Are you sad, Assaad?" I asked.

"No," he said. Part of me believed him, part of me didn't. I almost sensed a denial as we spoke. "Nervous about leaving the house," he said. "That's about it."

"What about the cat?" I asked, smiling.

It was still there, walking around — feral, a little tentative, seemingly more obese each time I glanced at it. It circled an empty tuna can that Assaad had set on the patio. "Nobody will take care of it," he told me. It was a typical line from Assaad, frustrated, discouraged, always about to give up. He lit a Winston and looked at the cat again. "I wish I had a wife like my cat."

He exhaled loudly and sat back in his chair.

"The only people I'm going to miss are you and Dr. Khairalla."

"That's it?" I asked.

He thought for a moment. "And the fat-tailed sheep, because we don't have fat-tailed sheep in the States."

He dangled his set of keys over the cat's head, jingling them. The cigarette hung from his mouth. He tried to get the cat to react, snapping his fingers, then shouting, "Come here!" The cat stayed still. I thought it might be irritated. Its back arched in a show of wariness and suspicion, it seemed only to want Assaad to leave it alone.

"Stupid cat," Assaad said finally.

Assaad's sense of not belonging drew me to him. Like him, I never felt a sense of community here, though I wanted to. After he left, I worried that my solitude was the legacy of families forever doomed to departures. I worried that, like Assaad, I would never really find home, not in Oklahoma, not in Maryland, not in Marjayoun. I suppose it is the curse of a generation always looking for something more, something better — the cost of too much freedom. Yet we search, sometimes without realizing it. I knew I wanted my own sense of *bayt*, and this is what had drawn me here. As spring settled in, I began to wonder whether I had already found it, as Bahija had.

A widow whose appearance did not betray her fifty years, Bahija was a wealthy woman, keeping the gold that Isber had left her tucked above the stained wood panels that ran the length of her roof. Dutifully, she raised the children who remained — Najib and Hoda. Considered old for a single daughter at twenty-five, Ratiba soon left, traveling overseas with her father's younger brother, Rashid.

Close in age to Isber, Rashid had always looked up to him. He had built his house next to Isber's, sharing a wall and chatting with him from their balconies, which opened onto Mount Hermon. After Isber died, Rashid had helped Bahija as much as he could, trying to serve as an authority figure and patriarch to his brother's children. But as the years passed and he began to feel his age, Rashid decided to join his sons Said and Kaleem, who had migrated to Brazil.

He bid farewell to Faris, his oldest brother, who chose to stay in Marjayoun, as a neighbor of Bahija's, and he sold his house to the Qurbans, relatives of his wife. As a gesture to his brother, Rashid made the offer to take Ratiba with him to Brazil, where she married the eldest of his sons, Said. She would never return.

In those years, only Nabeeh came back from America, arriving in 1931, more than a decade after he had emigrated with his sister Nabiha. Now in his thirties, he was eager to marry. Nabeeh had thought to stay a few months. He ended up staying nearly three years, his bedroom, next to the liwan, still bearing his name long after he returned to America. Bahija would have never confessed it, but Nabeeh, her oldest, was always her most beloved, and he returned her quiet, unspoken affection with respect that bordered on reverence.

Nabeeh traveled with his widowed mother, who had never ventured more than an hour's walk from Marjayoun. Their first stop was an obligation, the Convent of Our Lady of Saydnaya, an important site of Christian pilgrimage in the Middle East, perched on a rocky promontory in the mountains beyond Damascus. Renowned for its miracles of healing and renewal of faith, it was built by Emperor Justinian, whose Byzantine troops, fording the desert, struggled with thirst. In the distance, tradition has it, he saw a gazelle and gave chase. As he drew his bow, the gazelle transformed into a brilliant light, the Virgin Mary herself, who ordered that the emperor build a church to her on the spot — by legend, the place where Cain killed Abel. Justinian built the church. When he faced difficulties, Mary returned in his dream, again as a gazelle, with the plans for a convent. Miracles ensued.

As a two-year-old, an ill Nabeeh had faced death, and Bahija had prayed, promising to take her son to Saydnaya if he was healed. It was her own miracle. Thirty years later, she took him. Fresh from Oklahoma, an unruly and untamed frontier, one of whose towns Woody Guthrie would describe as the "shootingest, fist fightingest, bleedingest, gamblingest, gun, club and razor carryingest" places he had known, Nabeeh spent the night at the convent, amid frescoes illuminated by candles and behind stones reputed to have stood thirteen centuries, in a place inhabited since the Stone Age.

From there, they roamed Palestine for two weeks, visiting Nazareth, Jerusalem, Bethlehem, and the Cave of the Patriarchs in Hebron, before venturing back to Lebanon and a far more unfamiliar Beirut. Less than two hours by car, they returned again and again to Damascus, a more intimate city. There, a cousin, Kamal, had married. His wife played matchmaker, introducing Nabeeh to the parents of Adeeba al-Rayess, who was half his age. Beautiful, Nabeeh remembered thinking. But, he worried, "she's too young. What am I gonna do with her?"

Soon persuaded, he married her. At their wedding, on Christmas Eve, 1933, Bahija, never given to festivities, fretted that the dabke *they danced late into the night in the* liwan *would scratch, crack, or break her marble, which she had shined that day. Nabeeh and Adeeba, seven months pregnant with their first child, left in 1934, driven by a chauffeur so full of song that he honked his horn as percussion. Bahija suffered their departure in silence.*

16

SITARA

LIKE ISBER'S HOUSE, Marjayoun had its own secrets. So did every family. *Mastourin* was a word I had learned soon after I had arrived, at a lunch with Karim and his friends along the Hasbani River. Visiting was Bassima Eid, a lovely woman in her seventies, still strikingly beautiful, who mentioned the word to me. She was an expatriate now, living mostly in New York with her daughter and disdainful of convention. Even among fervent Christians, for whom devotion was a sign of social standing, she freely proclaimed that while she respected Christ, she had little time for talk of his divinity or the possibility of his miracles. She was no less contrarian about the civil war. "I miss it," she told me without irony. "I'm glad I went through it. It enriched us. We've seen things that people in the States would never dream of seeing."

The Americans who had become her neighbors lacked other familiar qualities. They were anything but *mastourin,* Bassima said. The word translated best as private, but it meant far more. Its root in Arabic means to cover, veil, hide, conceal, or disguise. With an apt pronoun, it can mean to shield, guard, or protect. There is a hint of pride in the description, perhaps ambition too. If someone is stuck in a bad marriage, she told me, no one else would know about it. No one complained about money, or confessed to failings, or admitted setbacks. To do so would reveal too much. They would remain *mastourin*. Interred in the earth or whispered behind a curtain, secrets were kept hidden.

So it was with those who participated in the diaspora of Marjayoun, settling in Oklahoma and Texas. Like old pictures stained with water, or wrinkled like the palm of a child's hand, they revealed only so much to the world. They spoke little of the past. There was always a *sitara*, a curtain, as they kept secrets buried in the customs and traditions they brought with them — all those Bedouin qualities of shame and honor.

By the time Raeefa arrived in America, a community had already gathered there. Stories of these immigrants have proliferated through the years. One Lebanese was put on the wrong ship in Marseilles. It was two years before he realized he was in Australia, not the United States. Nearly 6,000 came to America in 1907, and more than 9,000 in 1913 and 1914. The numbers dropped during World War I amid the ravages of the seferberlik, *the forced conscription, but then grew again, reaching a peak of 5,105 in 1921, the year after Raeefa and her siblings immigrated. Heeding a mood of hostility, infused with the xenophobia that haunts American history, restrictive legislation followed. The Quota Limit Act of May 1921 would allow only 882 Syrians to enter the United States. The Immigration Act of 1924 reduced the number to 100.*

Eid al-Khoury had been the first to leave Marjayoun, peddling his way from New York to Oklahoma, then as much a frontier as any locale in America. For many settlers, that meant promise, drawn in part by the lure of inexpensive land and the booming coal fields of the Choctaw Nation, where mining peaked around World War I, leaving behind a slew of towns named for the mine's owners and operators: Haileyville, Dow, Wilburton, Adamson, Alderson, and Phillips. Inaugurated by a well named the Nellie Johnson, with its iconic image of a black plume dwarfing its wooden scaffold, oil fueled communities in Oklahoma that boomed as quickly as they collapsed, leaving their inhabitants — in the words of Woody Guthrie, who lived in one such town — "busted, disgusted, and

*not to be trusted." Only a few of those Lebanese immigrants in those days
were miners and oil workers; most turned to business, opening groceries
and dry-goods stores that stitched the mercantile tapestry of the fledgling
towns for as long as they lasted. On arrival, George Shahdy even became
a postmaster, and in 1898, in a nostalgic gesture, he named the post office
Syria, Oklahoma Territory. Itinerant as those immigrants were, Syria
closed on July 31, 1907.*

Marjayoun gradually faded into memory.

Cecil had returned to Marjayoun in April for the first time since his rebuttal to my *Washington Post* piece was published in the town's magazine. By now, I had learned not to bring up the subject. Not that Cecil
would have been bashful about it or, God forbid, embarrassed. Indeed,
he seemed to take delight in mentioning to others in my presence how
much he disliked my essay on the town. It would forever haunt me, he
once declared, savoring his words.

To fellow guests, he insisted that I had visited in the middle of
winter, when Marjayoun slumbered, and that I had talked to too few
people, and those I talked to were ill informed. I had closed my eyes
to the potential and promise of the town, whose renaissance merely
awaited a true peace. I never confronted Cecil, someone I considered
a friend. I knew enough about respect these days to choose my battles. Cecil liked to say he was old enough to speak in whatever way he
wanted. I was young enough not to.

That month, I joined him on a Sunday for dinner at a place we had
nicknamed the Swiss Chalet, given its rustic look. At dinner, Cecil
grew irritated that the waiter had failed to bring the silverware right
away. "We're not Bedouins," he told the owner's wife, a buxom woman
whose beauty resisted age.

At times I lost track of our conversation: memories of his time in
Tunisia, the opening of his grandson's film in New York that Cecil
planned to attend, his daughter's very British husband, and the crisis — that ever-present crisis — in Lebanon, gathering more momen-
tum each day. "I call it the theater of the absurd," he told me.

All that was preamble, though, to a revelation about Dr. Khairalla, who, as mentioned earlier, had run the Marjayoun Hospital. Everyone in southern Lebanon, including Hezbollah's foes, describes as the Liberation the moment that the Israeli occupation ended in May 2000. With the Israelis gone, Hezbollah soon took over the hospital in Bint Jbeil. Amal, another Shiite movement, of far less discipline and far more corruption, seized Dr. Khairalla's hospital in Marjayoun. Soon after Amal's assumption of authority, Dr. Khairalla was formally accused of collaborating with the Israelis, and since he was still nominally an employee of the Lebanese Ministry of Health, he had to face a trial. (Cecil thought the entire episode was ludicrous. The Israelis had actually helped the hospital immensely; he thought they had probably spent $2 million a year or more on it. They had built a new maternity ward, another building, and a helicopter pad.)

The trial began in 2003, and Cecil discovered, undoubtedly surreptitiously, that the judge was Druze. The inveterate insider, Cecil went straight to the head of the Druze community, the mercurial Walid Jumblatt. "Walid was helpful," Cecil said simply. He still had that secretiveness of a government adviser — understanding that when knowledge is power, fewer words are better. "I was the only person who made any show of support for Dr. Khairalla," he told me. "He was extremely disappointed in the people of Jedeida. They did nothing for him. And he did so much." Cecil shook his head, in the tentative way that was by now familiar, as we both finished our beers, a local brand called Almaza. "He was treated very badly," he said again. "That's why he keeps to himself."

At night in Marjayoun, I gazed again and again at the old photos of those who had lived in or left Isber's house. One looks more like a painting, a fin-de-siècle rendering of Raeefa as a beautiful seventeen-year-old woman, seated in a carved wooden chair. Her short hair is a soft brown with reddish tones, the legacy of her father, Isber, memorable for the shade of his tossed locks. Her lips are folded into an expression that is less smile and more tenacity, at odds with her tiny, five-foot frame. Raeefa's pale features melt into her genteel clothes, a white dress and shawl that drape lazily over her body. There is an angelic quality to this portrait, painted in grays that threaten to blur. Though still a teenager, my grandmother appears dignified, serious. Life, in this moment, captured so fleetingly, seems calm.

Not as it had been.

After fording the Red River with her aunt and uncle, Raeefa, still just twelve years old, had met her brother Nabeeh in Wilson, Oklahoma, a town named for Charles Wilson, secretary to the circus magnate John Ringling, who helped build a railroad envisioned as stretching to the West Coast; it went no farther than Wilson.

Nabeeh had already left New York, where he worked at a dry-goods store, and as a peddler, and then, six days a week, at the Edison factory in New Jersey, building a product whose purpose he never knew. Tiring of the routine, Nabeeh had heard that his father's sister Khalaya was living in Oklahoma City, where her husband, Faris Tannous, had settled after peddling across the country. He decided to go, suspecting he would be even more welcome, since they had no children. In September 1920, six months after disembarking at Ellis Island, he arrived at 5 P.M. by train at the Santa Fe station, where fifty friends and relatives from Marjayoun had gathered to greet him. "Before I seen the people and before I knew what I am going to do," Nabeeh remembered thinking, "I just liked the looks of the city." With a loan of $200 from his aunt, Nabeeh and Faris, fond of drink and illiterate, opened a grocery store in Oklahoma City. For more than a year they ran it together, delivering goods to their customers by horseback before buying a car for $500.

Nabeeh brought Raeefa with him to Khalaya and Faris's house, and she soon enrolled in school. Everyone worked in those days, and Raeefa found herself spending afternoons at a cookie processing plant along Western Avenue. Diminutive as she was, she stood on a box so that she could pack the cookies in cartons. By nightfall, when she came home, Khalaya had left dirty clothes for her to wash. Whenever Raeefa was late, her aunt scolded her. "What were you doing?" she barked. "On the corner, looking for boys?"

Raeefa had been pampered, living a life befitting the daughter of an Ottoman country gentleman. No need went unanswered; no one but Isber had to work. In Oklahoma, her aunt forced Raeefa to stop school in eighth grade. She had learned enough. A gentle and encouraging charac-

ter, the principal, impressed with the girl's mind, tried to keep her. "If you would just let her stay, I'll pay all of her expenses," he offered the aunt. In a fit of old-world pride, a furious Khalaya scoffed at the offer and the gall of his intervention. Raeefa had learned enough, her aunt said again.

The conversation ended, as did communication with the earnest principal.

A few days later, Dr. Khairalla returned again from Beirut, where he was receiving treatment for his cancer. His face was tanned a rugged brown from spending hours in his garden, making for a vigorous contrast to his groomed gray hair. I felt he looked more vibrant, even revived, though I knew he wasn't. We were meeting for lunch, but as was his habit, he first sauntered through my garden, hardly acknowledging my presence, and inspected the plants we had bought together in Jibchit and the ones he had given me — a squat cactus, a passiflora, an emerging crape myrtle. I needed iron for the wisteria and fertilizer for the rest — nitrates for foliage, he said, "to grow more," and phosphates for roots, "to make them stronger." Potassium would help everything bloom, flower, and have bigger fruit. He grimaced as he glanced at a jasmine he had given me. It was little more than a sprig, and I had yet to plant it, looking for a place that I thought would properly acknowledge it as a gift from Dr. Khairalla. I was too embarrassed to tell him that, and in the time I had waited, it had withered into an autumnal auburn, its leaves shriveling in neglect.

Put it in the ground immediately, he told me, and water it every day. His voice was urgent. To almost everything he did he brought a doggedness, whether it was fashioning grafts of pear trees and coaxing an olive bonsai or building a wall for a reservoir and hoisting stones into a staircase that ran his garden's length. Each task was meditated upon and measured and plotted and figured and reconsidered, imbued with a respect that should define life. He could do nothing about his own dying body, decaying within, but he could master his garden in the spring, enticing all those timid roots, persuading the pruned branches, cajoling the many blossoms. The fate of the jasmine had hurt him;

I was careless in leaving it unattended, blasé about planting it. I had failed to appreciate what I could do.

We got in his car, and he told me he wanted me to hear a Lebanese violinist of Palestinian origin, whom he considered the best in the Arab world. His name was Abboud Abdel-Aal. "It's as if his violin is talking," Dr. Khairalla said to me. Abdel-Aal was playing "Al-Atlal," the ode by Oum Kalthoum that had made me understand why the price of hashish always went up in Cairo before she performed for hours, defying stamina. (Her voice possessed glass-shattering power; microphones were kept a half meter away from her mouth.) Dr. Khairalla looked through the windshield, lost in the lonesome chords that joined us, somehow plaintive yet bold. To this day, I don't remember our saying anything. Words would have intruded.

We sat at a table in a restaurant on the Boulevard called Road Runner. Both of us ordered pepper steak and Almaza beer. Cecil had told me that when Dr. Khairalla first learned he had cancer, three years before, he didn't tell his son and daughter. Burdened by a father's pride, he would not inflict his children with his worries. Perhaps shame played a part, too. As a doctor, he had always urged his patients to have a prostate exam, and he had not. I wanted to ask him at lunch about what Cecil had told me, that no one stood up for him after he was accused of collaboration. I knew I couldn't ask him directly, so we started talking about his long career in medicine. The path was never his choice, though I think he ultimately liked being a country doctor.

"The closer you go to Beirut," he told me, "the worse people become." I smiled. But what about the people here? I asked. Were they grateful? "Some appreciate," he said, "some are indifferent." To say more would have reflected arrogance, and Dr. Khairalla was unfailingly modest. In a town where swagger was often requisite, deference seen as weakness, and diffidence as stupidity, here was a man who insisted on humility. "In general," he said, "I think people respect me."

When another beer arrived, we began talking about his ordeal after the Israeli occupation ended, and he was more forthcoming than I thought he would be. Two or three days after the withdrawal, what he called *zoaran*—punks with guns—showed up at the hospital. They belonged to Amal, the unruly stepchild of Shiite politics. Turn the facility over to us, they insisted. Dr. Khairalla said he stood his ground. He told them he would accept orders only from the director-general of the Ministry of Health or from the minister himself, a man named Karam Karam, who was born in the neighboring village of Khiam. They relented, still making sense of the landscape that they had just inherited, but the next year proved a disaster for him.

Amal's enforcers kept returning, insisting that he hire as many as sixty of their own people. They staged inspections every so often. "They tortured me for a year," he said. Dr. Khairalla recounted the humiliations and disgraces with little emotion, but I could tell he was hurt. In the end, the government decided to privatize the hospital, and in a country where every deal provides for a percentage for some party, Amal seized control. Dr. Khairalla went home, collecting his salary for three more years, until he retired at age sixty-four. He was replaced by a doctor from nearby Blatt, a loyal if opportunistic follower of Nabih Berri, the speaker of parliament and leader of Amal, and the hospital began its slow, painful, and inevitable decline.

That was only the beginning of his ordeal, of course. The charges soon followed. They were ridiculous, he remembered, even by the standards of a kangaroo court. One charge claimed he was a captain in the South Lebanon Army militia, an accusation that bewildered him. Another said he was an adviser to General Antoine Lahad, the man who replaced Saad Haddad as the Israelis' man in south Lebanon, running the militia that they paid, armed, and trained to do their dirty work. He freely acknowledged that he went to Lahad often to resolve problems between the hospital and Lahad's men. Lahad was the only authority in town; visiting him was the only recourse. But conversations didn't make him a traitor.

Dr. Khairalla was eventually summoned to the court in Beirut three times. "For one year, I lived on my nerves," he told me. In the end he was convicted, but was not given any jail time, and he credited Cecil with helping. A month after the trial, he went with Ivanka to Kazakhstan for his son's wedding, then returned to Marjayoun, falling into the obscurity of illness and retirement, where I had met him.

I asked him if he was upset that Cecil was the only one to speak up for him during the ordeal. "Of course I'm angry," he blurted out. He typically hesitated in his answers, but not this time. His face was a grimace, and his words tumbled out. It was the first time I had seen him this aroused. Hoping to blunt his anger, I told him I thought people in Marjayoun were simply afraid. "They're not afraid," he said, "they're cowards. This is their mentality. They look solely after their interests."

As we drove back to my house, we listened again to Abboud Abdel-Aal. The music filled the car, and again we said hardly a word. His violin sounded like a *nay,* the Arabic flute — soft, with a hint of melancholy and a suggestion of loneliness. The same song played.

> *O sleepless one who slumbers and remembers the promise*
> *   when you wake up,*
> *Know that if a wound begins to recover, another crops up*
> *   with the memory,*
> *So learn to forget and learn to erase it.*
> *My darling, everything is fated.*
> *It is not by our hands that we make our misfortune.*

Halfway to my house, we drove past a sign for the Marjayoun Hospital. I thought only I had noticed it. But as it flitted past the car window, the doctor turned the volume up a little louder.

In time, Raeefa's aunt had welcomed to their home Jacob Rashid, another wealthy immigrant from Marjayoun living in Fort Madison, Iowa. The visitor was looking for a bride for his seventeen-year-old son, Faris, in part to stop him from marrying an American-born woman in Iowa. Raeefa's aunt and Jacob negotiated the arrangement, as Raeefa listened; Aunt Khalaya was in charge; she even took the engagement ring to a jeweler to determine the value of the diamond. Married soon, Raeefa left by train for Fort Madison, where Jacob opened a grocery store for his son and new daughter-in-law.

A few months later, when Raeefa was pregnant, Faris, too young to marry and too old to be so feckless, deserted her. Alone in the store,

Raeefa eventually had to sell it. She was too proud to rely on her husband's parents, and in the months after giving birth, Raeefa left her baby with her mother-in-law and began peddling on her own to earn money, carrying door-to-door a trunk almost her size filled with linens and threads. She managed to save about $900, a substantial sum then. With it she bought herself a remnant of her gilded life in Marjayoun, a fur coat and hat. "Buy one dress and make it a nice dress," she would tell her daughter, remembering those days.

The coat wasn't the only remnant of Marjayoun; Raeefa had inherited the temperament of Isber and Bahija as well. She suffered in silence. She chose her words; nothing was extraneous. Silence sometimes said more. Like them, she let others see what she wanted them to see. In time, others around her did. When Elva was still a baby, Raeefa posed for their portrait and sent the picture to her itinerant husband in hopes of drawing him back home, where his daughter awaited him. Soon, Faris did return. Back in Fort Madison, he and a friend took Raeefa to a club near the Mississippi River, and he presented divorce papers to her. He had married another woman, and bigamy constituted a crime. Still more of Marjayoun than America, Raeefa refused. Divorce was disgraceful, almost unheard of at home, and she would not bring shame on her family.

"If you don't sign it," Faris told her, his eyes wild and his friend there to add an intimidating edge, "we're going to take Elva and throw her in the river."

Raeefa signed.

Always affectionate toward Raeefa, her mother-in-law, Sadie, soon sat her down. They sipped cups of coffee, brewed as it was in Marjayoun, grounds settling on the bottom.

Raeefa was just nineteen. "You're young. You'll get married again. I don't want you to waste your life," said Sadie, who had given birth to seventeen children, eleven of whom lived. "Go and stay with your brother." Her father-in-law, Jacob, agreed, but insisted the child stay in Fort Madison. They would raise her better, as money for them was not a concern. Raeefa refused, and Sadie agreed. "I won't do that to her," Sadie told her husband. Raeefa offered a compromise: She would keep Elva and ask for nothing else. Grudgingly, Jacob relented, and Raeefa boarded the train from Fort Madison for the last time, rejoining Nabeeh in Oklahoma City, where the community of their countrymen was still lacking in many ways.

There was no Orthodox church there in those days. The Reverend Shukrallah Shadid, who was married to one of Isber Samara's sisters and who had traveled with the family of Miqbal and Abdullah Shadid aboard the Red Star Line in 1920, celebrated the divine liturgy from his home. He would do so for eleven years. In that evolving community, the reverend, known by all as Khoury Shukrallah, was the axis around which all revolved.

He soon visited Raeefa with a proposition. He wanted her to marry Abdullah Shadid, who had divorced his first wife. "He has a child and you have a child," he told her in Arabic. "I know he's a good man, and I want you to meet him. It would be good for you both to get married."

Khoury Shukrallah brought Abdullah to Nabeeh's house, and several weeks later they were married. As a new bride, Raeefa joined the traveling clan of Abdullah Shadid's family, a house of personalities as strong as hers.

17

SALTED MIQTA

D ID YOU SEE the *kanz,* the treasure?" George Jaradi asked me on a clear morning days before I was to move in down-stairs — or at least hoped to.
"Did you see the gold buried there?"

Next to the house was a cistern, a hole in the ground that, until now, had been concealed by a concrete slab about six inches thick, itself buried beneath dirt that was almost as deep. The cistern's sides were built of stone, stretching eight feet into the ground. Several inches of water gathered at the bottom. As I peered into it, the cistern looked like a grotto or an excavated tomb. It was big, revelatory, and, more than anything else, a mess.

I had heard stories about Bahija Abla watering her tomatoes from a spring in the garden. Several times a day, from morning to evening, she would fill her watering can with its long spout and douse the tomatoes. It was her routine, ordering the day like the call to prayer in neighbor-ing villages. I had suspected that the cistern had long ago caved in, or was lost when the road beside the house was widened a few years before. Now, as Toama's son helped me plant a plum tree, here it was, uncovered once again. But it was filled with shit, and Toama, George, and everyone else soon described it as such: *jourat al-khara,* the shit-hole. Albert Haddad, that menacing presence in the house for so long, had connected his sewage pipe to the cistern, making it a septic tank.

Some had experienced better luck. Shibil and my landlord, Michel Fardisi, had told me that during World War I and World War II, many fleeing villagers hid what money they had in their houses and yards. One such villager was Elie Bayoud, a relative of Shibil's by marriage. Repairing Bayoud's house, workers stumbled on three jars of gold coins behind a wall. With little discussion, they took it all. Or as Shibil put it, "They fled and never came back." It sounded apocryphal to me — who saw them find the gold? — but Shibil insisted it was true.

I, on the other hand, was left to dig shit out of the cistern. The plum tree planted, I begged two workers to help me and, at the right price, they agreed. We spread the stuff in the garden like peat moss. I had already started envisioning what I would plant in it — the squash and peppers that reminded me of my grandfather's garden on the harsh plains of the Texas Panhandle.

Soon after he had arrived in Texas, Raeefa's future husband, Abdullah Shadid, had become a peddler, traveling the roads of a sunburned countryside that smelled of cattle. Gold, his countrymen in Marjayoun predicted he would find in America. But fortunes were accumulated, not found, and many were built through his new trade. If there was one occupation familiar to every Lebanese family, it was peddling, a trade suitable for a country as yet unfamiliar with cars and buses and outlet malls. For those who worked hard, profits were good, far better than a factory worker's wage. Arab immigrants set out across America and to small towns in Mexico with suitcases filled with ready-to-wear clothes and bolts of cloth. They were a familiar sight in remote reaches of Brazil.

They sold everything from collar buttons, shoelaces, ribbons, handkerchiefs, lace, garters, and suspenders to exotic stuff from the Orient: handicrafts of olive wood and mother-of-pearl, embroidery from Syria, crucifixes, rosaries, and Orthodox icons that took on an air of authenticity when marketed by people from the Holy Land, even if the goods themselves were manufactured in New York. Most of the peddlers dragged trunks, a fortunate few had a horse and wagon, different from Isber Samara only in their goods and climes. They had traded the Houran for towns with names like Whitehorse, Driftwood, and Cherokee, all no more than a few miles from Syria, Oklahoma Territory.

Abdullah and his brother Miqbal lacked some of their countrymen's single-mindedness. Work was a means to enjoy life, and with a certain zest, humor, and volatility, both of them had, filling their evenings with song and dance, playfully pitting one brother's voice against the other's. Being younger, Abdullah was the more restless. Peddling gave him freedom, as he plodded down the roads with his sleeves rolled up and offered wives and their daughters pins, needles, and thread. He often flirted more than he sold. He soon sought work in the Texas oil fields before he was called up by the military, after which he brought his wife, Vera, a woman as tempestuous as he was temperamental, to Oklahoma. The marriage was volatile, their nights spent fighting then reconciling with an energetic dance to the scratchy records on a gramophone.

It didn't last. Vera soon left him, with their son Abdullah had named for his father, Ayyash, the friend of Isber Samara's in the Houran. Each moved on. The family never saw Vera again; she died a young woman, in her fifties. Abdullah wandered more, eventually joining the rest of his family — his mother Shawaqa, his sisters Nabeeha and Najiba, and his brother Hana — all of whom had left Beirut in 1920 on the Red Star Line with the Reverend Shukrallah Shadid.

"Is this the America you brought me to?" Shawaqa yelled at her son.

After his family arrived, Miqbal found them work on a farm owned by a Lebanese family near the western Oklahoma town of Brinkman, a prospering locale. At night, Abdullah and his newly arrived mother and siblings huddled in a wooden shack smaller than their stone home in Lebanon, itself modest. During the day they picked cotton. Bent at the waist, moving down row after irrigated row, they plucked one boll at a time, depositing them in a long and narrow sack stitched of white denim, bought from a nearby dry-goods store and slung over their shoulders with a wide strap. By sunset, their fingers bled, pricked by the sharp prongs that held the fluffy cotton snugly in the boll. Shawaqa's anger grew

*through the humid summer of the Oklahoma plains, where the horizon
ended only when eyesight surrendered.*

"Is this America?" she shouted again.

*Soon they would pack up their belongings and go to Detroit, though
Miqbal, who stayed behind, was leery. "Oh, you don't want to go to a big
town, it's dangerous," Nabeeha remembered him telling them. "But we
didn't listen to him, and we went anyway," she said. In the 1920s, Detroit
was booming, and anything was better than picking cotton. Hundreds of
men from historic Syria were already toiling in the automobile plants,
where they were known as Syrians or Turks, and Hana soon joined
them. Abdullah worked at a produce firm. Nabeeha and Najiba were
hired at the National Biscuit Company and wore white uniforms. All of
them lived together in a wood-frame two-story house with their ailing
mother, Shawaqa, who loathed the long midwestern winters. The family
lasted three years in Detroit. The end came on a cold, snowy, blustery
night in 1924, with too little heat in the house and not enough money to
keep them there.*

Abdullah's words were simple. "Let's go back to Oklahoma," he said.

*They did, to the same place Raeefa had found refuge. And they never
left again.*

Camille Salameh, the carpenter from neighboring Qlayaa and a man
so utterly lacking in punctuality that he measured time in seasons,
worked at the *warshe* for three days fastening doors to the frames of
four entrances. Fadi Ghabar, a Druze from Ibl al-Saqi, joined him,
cleaning more tile with his lumbering machine. But Camille didn't
want to finish until Fadi did. And Fadi didn't want to finish until Camille did. I wanted to clean the downstairs, but Toama wanted me to
wait on Camille (and Fadi). Emad Deeba came to install the water
heater, but told me to wait a day for the rest of the deliveries. The day
arrived, and they didn't. I learned some old Arabic words by necessity.
Bortash was one; it meant, effectively, a small slab of stone or marble
that serves as a doorstop. I came to know *bortash* only because I didn't
have one.

One decision prompted another, which brought another holdup, which demanded another decision, which delayed us more. So I did what I could. In the rooms Fadi had finished, I spent hours scrubbing the tile, still marked with paint, sometimes scraping it with a table knife. I cleaned the downstairs kitchen, bedroom, and bathroom — washing the fixtures, the floors, finding dirt in every corner and every crevice. I swept, mopped, then got on my hands and knees and took a cloth to the tile, roaming back and forth. I scrubbed the windows and shutters, fighting a losing battle against the dust and accumulated dirt of construction.

Every so often, Camille would leave his work on the doors, so behind schedule that the schedule meant nothing anymore, and come over to chat while I was on my hands and knees.

Camille loved conversation, be it about his prospects of obtaining an American visa or about how much better life in the south was under the Israeli occupation. The latter was a favorite refrain of Camille's. For most in these parts, the fondness was rarely for the Israelis themselves, but rather for the money they brought to Marjayoun. Even Shibil, the most militant of anyone I knew in his views of Israel, acknowledged the point. "You're still in a big *habis,* a big prison, but when the Jews were here, we lived better," he once told me. "They had money, and we had money. I was working. Electricity was free, water was free, no bills were paid, and you were making money. *Raah al-shekel wa ijit al-mashakil,*" he said, recalling a saying. "The shekel went, and the problems came." He paused. "Still, I wouldn't trade money for the Jews. I wouldn't trade occupation for my well-being. Definitely, I wouldn't."

Camille was less conflicted. "I look back on it as the best time Marjayoun has ever lived," he said.

A Maronite Catholic, Camille was the most vehement of the Christian workers when it came to Muslims. He didn't want me to hire them. In all the time since the occupation ended in 2000, what even he called the liberation, not one Muslim had offered him a job. So why should we hire them? he asked.

I usually listened to Camille's screeds whether I agreed with him or not. His politics were so far from mine as to sound like another language. But I also understood that Camille was a product of his environment: a village, Qlayaa, that prospered as it had never imagined under Israel's watch. Money was plentiful, residents had work, a sense of security reigned, and the Christians still acted as if they ran the place. Take the houses, he told me. Before the liberation, between ten and fifteen were built every year. In the seven years since, "only six houses were built — and they're not even finished."

In a moment, in other words, it had all ended. Many of his friends fled — for Israel or anywhere they could snare a visa. Salaries evaporated. To him, the Christians were dying, soon to be extinct in a country that, in his words, is "the worst state in the world.

"I have no hope," he said.

A friend of mine once recounted a conversation she had with Camille as he drove her to Nabatiyeh. Eighty percent of his friends lived in Israel, he acknowledged. They drove down a winding road next to the Israeli town of Metulla, perched on a hill and facing Marjayoun. He stopped the car to point out a green house. "That's where one of my friends lives," he said. "Can you believe how close we are to him? And yet I haven't seen him for a decade, almost. Aren't you wondering how I know he lives here? I know because, right after the liberation, we used to go right up to his house and hang out together. We'd hang out and no one would say anything. We can't do that anymore."

Camille was staring at Metulla's brightly painted houses, only a few steps away, past some shrubbery. "Look how beautiful Israel is," he said. When she asked if he would live there if he had the opportunity, he jokingly swerved the car off the road toward Metulla. "I'd go there right now if I could," he said. "Lebanon is beautiful, okay, but we can't live here. Do you know how they live there? They have jobs, social security. The handicapped have rights there. They're happy. We just want to live, and over there, you can."

An academic study written soon after World War II at Columbia University assessed the "Tentative Formulation of Some Prominent Aspects of Syrian Culture."

The study was as forgettable as one passage was memorable, a mix of remarkable chauvinism and perhaps a grain of truth concerning the immigrants' desire to adapt. "Let us take a Maronite from a small village in Lebanon, and see how he would act in interpersonal relationships," it said. "In dealing with a religious American Protestant, our Maronite would think of himself and talk of himself as originating from the Holy Land; with an American Catholic, the Maronite would think of himself as a Catholic; with a Maronite, he is a Maronite. Dealing with a Protestant from Beirut, he is a Christian Lebanese, but meeting a person from Aleppo or Damascus, he is a Syrian. Making the acquaintance of a Moslem from Egypt, he thinks of himself as an Arab, while in talking to a sophisticated American, he is a Phoenician. Applying for a job in a Jewish firm, he is a Semite, but if the firm is a religious Christian group of any sect, then evidently he is a Christian. In dealing with governmental and patriotic organizations, he is one hundred per-cent American."

Like the Maronite, the Samaras, Shadids, and other Lebanese from Marjayoun were determined to be like their neighbors in Oklahoma. English was one way, but their tongues never quite negotiated it. Names were another way. Some were bestowed quite randomly at Ellis Island by American officials. Other Americanizations of Lebanese names came in other ways. A Lebanese woman named Mary, who arrived in Maine around Christmastime, was frequently greeted with "Merry Christmas." How kind they are to greet me by name, she thought. The eventual explanation did little to temper her glee. From then on, she decided her name would be Mary Christmas.

The Shadids looked for names that had at least some resemblance to their Arabic roots. Miqbal became Mack, and Abdullah, Albert. Hana took the Anglicized name of John. Nabeeha reached further. She was known as Nevada, and she called her sister Bee. Last names underwent a transformation as well. Shadid, pronounced in Arabic as Shdeed, only rarely kept its original. Miqbal's family changed it to SHAY-did. Abdullah's went by SHA-did. At least those names were spelled the same. Naifehs, by choice or the whim of an uncomprehending immigration official, became Naify, Nayphe, Nayfa, Naffah, or Knafe. There were Kouris, Khourys, Courys, and Elkouris. Harroz became Harris; Barakat, Bar-

kett; and Dabaghi, Debakey. One of the best known of the Marjayounis to arrive in Oklahoma was Adeeb Eid, and he was rare in choosing his own name. As a teenager he had become fascinated with baseball, his idol Babe Ruth. Adeeb became Babe, and Eid became Eddie. In years to come, no one would remember the Arabic, only Babe Eddie.

On the occasions they adhered to tradition, problems sometimes followed. Isber's oldest son, Nabeeh, named his son for his father, as was the custom. When he received the birth certificate from the Oklahoma City hospital, it read, "Ester Samara, female."

"Hell, this is no Ester!" Nabeeh exclaimed. "He is Isber, and he is male!"

As accommodating as they might have been with their names, they still clung defiantly to the traditions that they believed set them apart.

For all, food was the mainstay, whether it was the lavish lunch that followed a baptism — usually in a sturdy round tub in the days before the church was built — or the more ordinary sharing figs and miqta, with which salt was always required. Abdullah would even plant a small patch of wheat, that crop first cultivated in the Fertile Crescent, to make a dish called frike, best served with lamb. Their native-born visitors expressed surprise at some of the rituals. At the sumptuous meal that followed one baptism, invariably conducted by the Reverend Shukrallah Shadid, a man named Mr. Maddox watched the Lebanese eat khibiz marqooq, a bread thin as paper and folded in half.

"My God," he said to his wife, "they're eating their napkins!"

On another occasion, an American couple put the bread in their laps.

To the immigrants' American neighbors, weddings were an endless source of amazement, the enthusiasm they inspired often finding its way into the pages of the Daily Oklahoman. An article in 1924, covering one wedding, spoke of a Syrian priest "chanting with nasal intonation peculiar to all worshippers of the East" and of "an almost Bacchanalian feast" that followed the ceremony. Another wedding, in 1913, between a couple from the Samara family, found itself on page two of the newspaper. Hearing "the din and uproar" of the reception, neighbors "within a block of the little shanty" called the police, who themselves feared that a "riot was in full swing." "We jes' have a leetle fun, das all," one of the participants *explained, in a reporter's rendering of his Arabic-accented English. The reporter himself seemed taken by the spectacle of the festivities.*

Socializing was a necessary priority. Visits kept to a village cadence; gossip was often dressed as concern. Conversations sometimes turned into hands of whist and games of poker, with walnuts used as chips. Impromptu parties were convened on any night. All that was needed was the derbake, *an hourglass-shaped drum that some say was invented before the chair. Abdullah Shadid liked to cup one hand around his ear as he crooned, his voice redolent of the three packs of cigarettes he smoked each day. The line of* dabke *soon formed, sometimes performed with such zeal and passion that the weaker would fall into their chairs, their legs still flopping beneath them, hewing to the dance's rhythm. Never a word of English was spoken. On those nights, they were back in Marjayoun. They were home, together.*

Finally, the day had arrived. In a scene recalling the Dust Bowl exodus from Oklahoma, I packed everything I had in Marjayoun in the back of a small Toyota pickup that belonged to Toama's brother-in-law. There it all was, the conglomeration of nine months of living in the town — two chairs, a couch, a dining table, its four chairs, a wardrobe, a desk, its chair, a refrigerator, an *ujaa* and its pipe, a few assorted tables, my clothes, and everything from the kitchen, from the last jars of olives and anise for a wintertime tea to the plates and silverware donated by a journalist who had departed Beirut and left behind what she couldn't carry. It all threatened to spill out the back of the truck, where Tolama's son, Ali, sat assuredly. "Two kilos of fish!" he shouted, in his best rendition of the fishmongers who ply Marjayoun's streets, water sloshing in their pickups turned aquariums. *Sultan Ibrahim!*

At Isber's, we disgorged everything into the Cave, the room that was closest to being completed. Since Camille, predictably, had not finished the doors, we propped the bed over the entrance outside. We grabbed an old door from Abu Jean's pile of detritus, swept it with a broom, and lodged it against the entrance inside. One light worked. I had yet to get bulbs for the others. There was no heat, and the April nights were chilly. As dogs barked outside, I wondered at one point how steadfast the doors really were. I quickly shook my head. It didn't really matter; I was too tired to worry. I put the mattress on the floor, piled as many blankets as I had on top, donned long underwear and a sweater, and crawled underneath.

For the first time in more than forty years, nearly a century after Isber Samara had begun building his legacy, someone in the family slept again in the house.

The next morning was a Sunday, not a workday in Marjayoun. As I walked around the house, I started to see beauty that I never expected. The tile was restored and rejuvenated. Shades of brown dominated the house, but provided accents of blue and red, always subtle. The paint job, by Joseph Abu Kheir, the son of an itinerant soldier, was quite satisfactory, but the stone, seemingly illuminated, was the centerpiece. I stared at the size of the stones and at their color, noticing imperfections everywhere, beautiful imperfections. The wall was built of row after row of stone, eleven high. It was barricade, foundation, backdrop, and entrance. Monumental in stature, they sprawled, unmovable, a part of the house that could not be changed and could not lose its identity. That wall, I suspected, could never be destroyed. Nor could the pointed arches before it, side by side, like Atlas's shoulders.

A truly Levantine innovation, the pointed arch's origins go back to ancient Mesopotamia, and the design was still commonly built in the Middle East when Muslim conquerors arrived millennia later. They, in turn, spread the pointed arch westward through the Mediterranean. As I stared, I thought back to Abu Salim Haddad, the last of the region's true stonemasons, who had adored arches. "Look at this arch, you could put all kinds of floors, five more floors on top of this building," he told me when we started. "The most important thing in a house is the arch. An arch won't fall down." I didn't understand then, but I did now. Survivors, these arches were now bearing something they never built for. What was once closed had become open, what was utilitarian had become lithe, in an elegance that happened as much by accident as by plan.

"You could pay a million dollars and you couldn't build stone like this," Toama said when he and Abu Jean walked through the house with me.

Abu Jean nodded in agreement, rare when Toama spoke. "Where are the *maalimeen* today to build it? None of them are left."

As they talked, I opened the shutters. What sun there was came in. It was muted, far softer than it had been when the house had been a *warshe,* open to the elements. But it bathed the floor and walls gracefully, catching the shades of the stone.

That afternoon, Cecil came over. In a measured gait, he had walked to my house. He was slightly stooped, and his shoulders were a bit hunched over. Earthy in their colors, his clothes were Jedeidani to the core: faded slacks, a green sweater, and a brownish gray, frayed jacket. I had not talked to Cecil in a couple of weeks, and he was eager to see what the house had become. "My advice to you is not to get involved in the Samara house," I had remembered him writing all those months ago, and now here it was, finally taking shape.

"It's a revelation, this house," he said. I had never heard him speak that way, his usually measured words taking on a sense of awe. "I never imagined it like this."

Cecil entered each room as if preparing a report on the place. In some, he hovered in silence. You could see that old brilliant mind of his still calculating the angles.

"When you first saw this house, you couldn't imagine this, could you?" he asked me.

By the time Raeefa married Abdullah, the number of Lebanese in Oklahoma had grown. There were more than seven hundred in those days — among them the Samaras and Shadids. The exodus from Marjayoun to America that began in the 1890s was now coming to an end, and their settlements in Oklahoma were becoming permanent.

With that permanence came danger and hardship. The family of Abdullah's brother Miqbal settled in Brinkman and opened a dry-goods store, cluttered with yellowing signs that declared a sale for "today only."

In their town, no one knew quite what to make of Miqbal's clan, which soon grew to ten children. The local boys bullied George, their oldest son, who spoke only Arabic when he arrived at school. In church, the Shadid children were harangued by a preacher who deemed them too dark to be Christian. "You're gonna go to hell if you're not saved!" he shouted at the girls. They ran home crying, telling their mother, Hafitha, of their fate.

"Don't worry," she told them, her English still too tentative to pronounce r's and never good enough to distinguish a p from a b. "You've already been babtized."

Hafitha shook her head, speaking softly. "You're not going to hell."

The Ku Klux Klan had revived in Oklahoma in those days, and violence, from race riots to lynchings, swept across the state like a scythe as the group targeted everyone from wayward Protestants to communities of African Americans, Jews, Native Americans, and Catholics. Their attacks, and those blamed on them, became so pronounced that Jack C. Walton, the state's progressive governor, declared martial law in two counties, and eventually in all of Oklahoma. (His enmity toward the Klan contributed to his eventual impeachment, after serving just ten months.)

Brinkman, on the Texas border, was not spared. At first, the threats to Miqbal and his family were communicated by whispers, the small talk of customers warning the gentle, amiable storekeeper, who made a point of speaking to customers in their own language—German, Spanish, Arabic, or English. Then they were conveyed more forcefully by townspeople who, in principle, refused to frequent the store. Occasionally, the police, whose ranks were filled with men who belonged to the Ku Klux Klan or sympathized with it, bluntly delivered the warning: They should leave. Miqbal's oldest child, Gladys, would never forget one night in 1923. In their front yard, near a small vegetable patch, a wooden cross was doused in gasoline and burned, its flames pricking the night air for hours. Since this was America, Miqbal insisted, he would ask the governor to act. Indeed, he dialed Walton's office, pleading for him to intervene. His office did. Miqbal's family never knew quite what happened—who was called, who was summoned, or who was warned. But no cross was seen again, nor was a threat whispered.

For insurance, Miqbal bought a pit bull he kept at the store. Rex, he named it.

Miqbal's charm seemed to court hardship, a circumstance perhaps attributable to the presence of the rather difficult Rahija, Miqbal's mother-in-law, who thrived on discord and chaos. With her daughter, Rahija would encourage fights with her son-in-law, then sit back to enjoy the spectacle, a sadistic smile spreading across her face. Births, in the event the baby was a girl, were disappointments. She raged at the coming of Gladys. Because she had wanted a boy, she poured acid on the breast of her daughter, making it impossible for her to nurse. When Rahija was away, Hafitha put sugar balls in her baby's mouth. It was not enough, and Gladys withered. Her chest burned, her spirit broken, Hafitha finally asked her sister Martha to nurse her infant daughter.

"Don't tell our mother," Hafitha begged her sister.

When Rahija died, her daughter's words were remembered.

"Don't cry for her," Hafitha said.

No one did. Rahija's son Frank, abandoned by his mother at six months in Lebanon, assumed her place as head of the family. To his sisters, as timid as they were gentle, Frank acted the way a son of Marjayoun would. Imperious, with a severe face and a scar over his left temple, his word was final. To his sick nephew, George, lying in bed and attended by a doctor, Frank arrived to prescribe his own treatment: horse urine. (Somehow, it worked.) When another of Miqbal's sons, Junior, was stricken with pneumonia, Frank dismissed the doctor and applied his own mustard plasters — mustard powder mixed with flour and water, thought to help bronchial problems — to the boy's chest. Just two years old, Junior died anyway, as Frank hovered over his frail body, shouting orders to Miqbal and his wife and treating without knowledge. As others watched, Frank put his hands over the boy's eyelids, closing them for the last time.

"He's gone," Frank said, without emotion.

With her golden brown hair, Pauline was the most beautiful of Miqbal and Hafitha's children, inheriting her father's charm. She was the fourth of ten siblings and had an independent streak. At twenty-one, she fell in love. It was not an adolescent crush. Both were adults, and both wanted *to spend their lives together. Frank fulminated and fumed, raved and ranted. He insisted that his niece marry a Lebanese man. Pauline was told that she could not date her neighbor in Brinkman.*

Hafitha toiled long hours in the kitchen; to her children it seemed she was always cooking. There, she had a bottle of carbolic acid, used to disinfect chickens raised in a shed beside the house. Soon after Frank delivered his order, Pauline drank the acid, a foul, corrosive liquid that causes stupor, deadens sensation, slows the pulse, then causes death. Pauline passed away before anyone knew what she had done.

Six weeks later, her admirer visited her brother David. He wanted to take a drive, anywhere, past the flat farms stretching to the horizon. David agreed out of sympathy, and the men piled into the family's 1935 blue Buick. After an hour or so they stopped the car and got out to survey the vista. While David's head was turned, Pauline's admirer walked from the door to the fender of the car. As David idly stared into the distance, the other man pulled out his own bottle of carbolic acid. He drank as she had. As was his wish, he died as she did.

"My identity is being reformulated," said George Dabbaghi, the son of my neighbor Maurice, the principal of Marjayoun National College. I didn't understand what he meant, so I asked him to explain. "The question is, do you keep living in the past, or do you look somewhere else? And if it's somewhere else, where is it?"

George and I had taken advantage of the spring weather to go on what we both termed a field trip. As garrulous as he was interesting, George acted as a guide of sorts. "There's a tremendous amount of history here," George said as we embarked in a rental car. "You have history under your eyes." He looked out at the landscape, as did I. There was the valley, then Mount Hermon, just fingers of snow remaining, barely tracing the mountain's contours, an escarpment over which he said invaders had been coming for four thousand years.

We drove through the *marj* below the town. I had never taken the route we chose, and the beauty and age of the place struck me. The British had fought the Vichy French here in World War II, and along a cratered road we crept past barriers of stone the British had built to block the passage of tanks. But what was so remarkable was the natural landscape. Olive trees grew over and around the old trenches. The fields of wheat were verdant. Where there had been figs, now olives and flowers grew. One flower was known as *showk al-jamal,* thorns of the camel, named for the animals that consumed it. As we climbed to the top of a flat hill, a possible Canaanite or Jewish burial ground, we surveyed the land, broken by borders on each side — Israel to the south, Jordan to the southeast, Syria to the east — none of which existed in Isber Samara's time.

As we stood there an elderly farmer came over, Abu Ali Wansa, from Dibin. He confessed that he had spent six months in a Lebanese prison after the liberation, presumably for collaborating with Israel, but he was more interested in regaling George and me with stories about how the various peoples in the region once got along well. They had intermarried, they were all tied to the land, and they had stuck to principles of coexistence rather than dogmatism in faith. I asked him what had happened. He pushed his cane forward, tapping the weedy ground, and spoke with a sense of the obvious.

"Wars," he said succinctly.

I thought about politics much the way George did, and as we stood there, we started talking about Arab Christendom. He was no chauvinist. There was none of Camille the carpenter's hostility. But like me, he was filled with a sense of fear and loss over what was happening to the Christians in Marjayoun, as the town withered. "I don't have the guts," George said, "to say goodbye and go." Thoughts of my grandparents flashed through my mind, their courage to leave. "We're never going to be in a position of making power in Lebanon," he said.

I had to agree with him. As Christians, we faced marginalization. We were not included in decision-making. To persist in that identity, we faced our own extinction. It reminded me of words that Assem Salam, an architect and friend of Cecil's in Beirut, had uttered during the war in 2006. What each community lacks, he said, are "the guarantees of its survival."

Power meant survival; without power, nothing endured. That was the arithmetic of the Middle East that had evolved from my Levant. A lack of power meant obsolescence, communities that were too small.

Take the fate of Christians in Iraq: "They're finished. They're gone," George said. He pointed at the hill where we stood with Abu Ali, the farmer. "That guy said we're all one people, but nobody is going to listen to you," he said. "You have to find identity somewhere else." And by identity, I thought he meant a notion of emotional or mental sovereignty. By identity, I thought he meant survival, or the power to imagine something more, something broader, another kind of community. "We're not as attached to the land as someone else might be."

We had lunch at his father's house afterward, and as conversation often did, it turned to the history of Marjayoun. The ideas echoed Cecil's talks with me, the way that Marjayoun once brought together religious communities. It was in part nostalgia, of course, but there was a truth in it, too. The notion spoke of power and confidence. It denoted identity that was less rigid. It took place in a land freer of borders, and there might even have been guarantees.

All this was gone now, my neighbor Maurice Dabbaghi said.

"The lost past," he called it.

18

PASSING DANGER

I WAS WAITING FOR Hikmat to visit on a spring night. For a week or so I had spent most of every evening outside, eating fresh almonds, sipping scotch, and feeling a peace that I had not felt in a long time. The house was utterly tranquil but for the sound of the wind blowing through the trees. The fragrance of jasmine enveloped me.

As I sat there, the porch lights highlighted the virile green of the plum tree and the silvery green of the two olives. I could see the faint outlines of the plants in the garden itself—the basil, the *rozana*, and the passiflora that Dr. Khairalla had given me. The breeze was warm. I had finally escaped from war. The reverberations of explosions and the noise of helicopters were all but gone. Since the death of Imad Mughniyeh in February, Lebanon's crisis had settled into the grimmest of stalemates. No one was fighting but, of course, no one was compromising, either. The specter of bloodshed continued to haunt the country. Television ads appealed for dialogue: "If not for us, then for our children: Talk to each other." Signs in stores declared that it was forbidden to talk politics inside. Yet in the streets, where knots of soldiers were deployed at intersections, there was fear.

To Mohammed Heidar, a Shiite from neighboring Dibin, Hikmat had once warned, "If anything happens, collect your children and leave Dibin on the spot. This time there's going to be no joke with you Shiites and Hezbollah." He meant the next war was going to be a showdown. To his wife, still afraid for her daughter, he was more reassuring. Wars happen in the summer, he said confidently, and it was still spring. "You don't want to send a soldier to die in the mud and the cold." Even if there was a war, we both insisted, Hezbollah could never be defeated. Nearly everyone seemed to agree on this point. The government, claiming itself the heir of Rafik Hariri, enjoyed the support of probably half the country, and importantly, the United States. But Hezbollah represented most of the Shia, and they were Lebanon's single largest group. Besides, its patrons, Iran and Syria, knew the country far better than the Americans.

Hikmat soon arrived, his first visit since I had moved in downstairs. Like a *maalim*, he walked through the house, shoulders thrown back, head cocked, nodding in approval. Perhaps more than anyone else in the town, Hikmat understood what the house meant to me.

"Now you have a relationship with this house," he told me. "Those pieces of shit in the States have nothing. Everything, everything, the door, the tile, the stone," he said, pointing. "You have ties with it. You know what it was. The simple act of discovering the arch, you did it. The simplest thing in the house, you have a relationship with it.

"It's part of your body now. It's the womb of your body. It's you, *khalas*," Hikmat said.

He looked at me seriously, like a father. Laila should come here, he instructed, and she should feel the house is hers. She should understand her relationship to the town, and yours. The line of continuity, from generation to generation, should not be broken again.

I poured him a scotch from the decanter he had given me, and we sat on the art deco furniture, upholstered in a wine-red fabric, that I had bought at a used-furniture market in the Beirut neighborhood of Basta. As he drank, he grew more reflective and personal.

"You can see into someone's heart when they're sleeping," he told me. He peered into his daughter's heart when she slept in her crib, he said. He saw into the heart of his wife, Amina, and sensed that she was good. For both, he said, he felt that God had set him on this earth to take care of them and protect them. "What's written on your forehead you'll see with your eyes," he told me, a saying I had heard often in Baghdad, meant to explain the inevitability of fate.

I could see the bond growing between Hikmat and his daughter, Miana, and part of me felt that it was the last element of his persona as a *zaim* — a father and protector — to emerge. He was sincere, too. "From now on, when I'm away from her, it kills me. I miss holding her. She loves me holding her." He sipped more scotch. "I was never scared of death," he said. "But now I don't want to die."

Khalil Abou Mrad, a friend of Dr. Khairalla's, had once told me a story. He was visiting a store selling roasted nuts when the vendor asked him how many children he had. "A daughter," Khalil replied. "Fuck that!" the guy shouted, a voice from the old world. "Is it worth all those years of marriage for one daughter?" To Hikmat's credit, I never heard him express regret at not having a son, and the way he talked about Miana made me jealous of not having my own daughter with me. I called her almost every day. I saw her for a few weeks every couple of months. But it wasn't enough, and I felt ashamed at not being with her more often. I told Hikmat that.

When I'm with her, I went on, she tells me after we read a bed-time story that I am "the best daddy in the whole wide world. There is no one better, ever." She means it, and she wants to hug me after she says it. My confession seemed to fill Hikmat with emotion. Maybe it was pity; maybe he sensed how much I needed, after all my absences, Laila's reassurances. But I suspect it was also his understanding of the bond between parent and child. As I talked, Hikmat's eyes welled up, and a tear fell down his cheek. He did nothing to remove it — either he didn't care or he was too embarrassed to draw attention to it. We were two fathers, both insecure, both worried about what would happen when our daughters became more aware, both terrified of the ways of the world and the future.

As we talked, Amina called. She asked Hikmat when he was coming home.

We had planned to finish the bottom floor of the house in January. We did it in April, and I shuddered at all that was left to be done. We had two months to complete the rest, and it seemed impossible.

One afternoon, after everyone headed home for lunch, I wandered up to the veranda above the stairs, known in Arabic as the *istayha*. From one corner, past a stone wall, a thicket of pomegranates, towering pine trees, and squat figs that spilled beyond them, I had a mesmerizing view of Mount Hermon. Cousins had told me that Bahija used to sit in the same corner doing needlework on her pillows and blankets. As I stood in that spot, I wondered what she would think of the house where she had presided for half a century.

An old piece of marble from the kitchen leaned against a stack of tiles on the veranda, along with fragments of a bygone doorstop. Two old windows were stacked against the dining room wall. Next to them was a resilient shutter whose color, the original green of the house's windows, I had always found so lovely. Remnants of Bahija's house were everywhere.

Bahija's autumn was Marjayoun's. No one disputed the town's decline, but she would have had little to say about the changes. As an elderly woman, she hardly ever left the house. Why would she? It had become her world. The cushions everywhere she had made, as she had made the drapes. She was partial to white; it seemed clean, for her the cardinal virtue. But for accents she sometimes dipped fabrics in tea, creating a beige, never too brown.

The outside world held few attractions for Bahija. Everyone she knew best remained inside the rooms of Isber's house. In its favored spot, the cherished portrait of Archbishop Elia Diab still hung, a lonely artifact of times gone by. Isber had often gazed at the image of his friend with fond-

ness. *The clergyman, one of the many who had fled, had bestowed that meaningful farewell before leaving town. "Remember me," read the inscription. Perhaps these were the words that Bahija lived by, as Isber was everywhere — on the balcony, or at his desk, or smoking his* nargileh *in the* liwan. *Bahija's children were everywhere, too. She could, she once told one of her daughters, still hear their voices in the halls. Did she hear the rhymes she had taught them as children?*

> *Oh Laila, oh Laila*
> *Oh Laila, there are no eyes like her eyes*
> *And the magic in her eyes,*
> *Oh Laila, oh Laila.*

Once the days had begun with her waiting for the Bedouin women to bring the bread. So many times she had promised Raeefa and Ratiba sandwiches of butter and sugar when the bread arrived. Now they were gone across the world, their lives echoed in letters alone.

Though she sometimes forgot whether she had put lemon leaves in the jars of olives she prepared in the fall, Bahija was still of sound mind during the years of World War II and after. She was respected in the town as a kind and dignified woman. She got up at 5 A.M. and brewed coffee, enjoying it on the istayha, *looking out at the hydrangeas flowering in white, blue, and purple. Sitting there, observed by her neighbors, she seemed younger than her years. Often she would take a break from her morning cleaning to chat with her nieces Laurinda and Wadad Abla, who lived downstairs with their husbands, or to talk to one of the soldiers from the nearby base. The house was always spotless. Like Isber, she was thrifty, sparing with the money that her husband left and that Raeefa and Nabeeh would send her. But she spent generously on her home. Like her husband, she would ask question after question until she settled on the best purchase. She stared at items for minutes as impatient vendors looked on.*

During the day, Halim Sukarieh would bring Bahija vegetables, fruit, and meat. Her gardener, Ali, would show up in the late morning, assisting her with the plants, tending her pomegranates and olives, helping her pick the parsley and mint, watering her vegetables as she grew older.

He was a striking man — one eye a lighter blue than the other. Hussein, in his black kaffiyeh, would bring her coal and work as a handyman, promptly fixing whatever was broken. She cooked elaborate lunches for all who came. By late afternoon she sat down to crochet, knit, and embroider. She had created almost everything around her, even the tablecloths, exquisitely rendered. For Christmas she would knit blouses for the children of her favorite relatives.

At sunset she lit the kerosene lamps, whose glass covers she scrubbed each morning. By 7 P.M., sometimes earlier, she prepared for sleep. She had moved into Nabeeh's room, reluctantly admitting to herself that he would never return. Her old bedroom she rented out to officers from the Marjayoun base. Most stayed for a year or two before they were transferred elsewhere. The soldiers became her helpers and guardians as she aged. But still, she lived alone. As the lamp cooled before sleep, she was no longer lulled by the whispers of children or her husband's footsteps in another room as she protected his family from danger.

Finally, the danger had passed, along with so much of Bahija's life.

Of the men who gathered at the *warshe* all those months, Abu Jassim was one of my favorites, even before I learned his name.

Curly-haired and stocky, with a smile that suggested grief, Abu Jassim loved to parade his English vocabulary, comprising about ten words, constantly rearranged and delivered in nonsensical combinations. "Here, three, door, mish good," he would tell me. I smiled, then spoke to him in Arabic: "How are you?" "Good," he would reply.

He was a perpetual skeptic, and often I caught a knowing glint in his eye as he watched the bravado of the Lebanese *maalimeen* in the *warshe*. Nevertheless, one day in May he surprised me by suggesting that we might just possibly finish the rest of the house. His optimism buoyed me, since at the moment the upstairs looked like a disassembled car, with its hundreds of engine parts spread out on the driveway. No one appeared interested in putting them back together. "Clutter" suggests untidiness. This was chaos, and it started at the foot of the driveway, where chunks of tar, remnants of window frames, and chipped and hammered stone gathered as a welcome.

But even I had to admit that Abu Jassim was right: much had been done here. Despite my perpetual frustration at the looming deadline, I imagined Isber pacing through this old house. I thought of all the original stonemasons and *maalimeen* hoisting the boulders, laying the tile, and buttressing the balconies, girded in rusted but exquisite *darabzin*. I wondered whether Isber shared my exasperation over the delays and disappointments. I looked at the stones throughout the house, now cleaned, with new mortar of the softest cream between them. We had decided to reveal the stones in the dining room and the *liwan* rather than covering them in cement. It was an innovation of sorts, my innovation: The stones were never meant to be exposed, but they felt warm to me, bringing forth a beauty as grand as the marble floor or the triple arches. Isber would have appreciated the touch. As I walked through the *liwan* and the dining room, I knew the pride he must have felt. My grandmother, only twelve when she left, would never have felt this way about the house. How could she, after all? She built her home elsewhere.

There is something uniform about the pictures of the dry-goods stores that Raeefa, Abdullah, their relatives, and other Lebanese ran in Oklahoma outposts like Texola, Lindsay, Brinkman, Snyder, and Sayre, some of them soon to wither into ghost towns. Gray backgrounds turning to yellow, one picture is almost indistinguishable from the next. The floors of the stores are always wood or concrete, and the counters, at least one row fashioned of glass, invariably run along the walls. Stacked on shelves near the ceiling are suitcases, and below them are bolts of fabric, tidily bundled textiles, boxes of shoes, cosmetics, thread, needles, and so on. Ready-to-wear clothing hangs from racks, and coats with fur collars are draped over a table. The store that Abdullah's brother Miqbal ran was adorned with advertisements only he could fashion, each bearing the lettering he was so proud of. These Prices Fairly Yell, *read one.*

Nearly every one of those immigrants was imbued with confidence. It was the cunning, the shatara, *of Marjayoun, honed by the determination of a person who has surrendered everything—family and home—for the unknown. Nothing was given, nothing was assured. The immigrants would work longer hours and prove to be more clever, even if it sometimes came with a sly wink. One enterprising Lebanese merchant was said to have left a broom by the counter. He charged each customer for it. If the customer returned angry, the cashier pointed to the broom.*

"Well, you bought it. Why did you leave it behind?" he asked.

· 19 ·

HOME

PERCHED ABOVE THE Saha was another old house, unin-
habited. It was a mansion, bigger than Isber's place, and if you
drove down the Boulevard you couldn't miss it. I always saw a
lesson in it: Here was the fate of Isber's house had we not begun work
on it.

I had gone to that effaced ruin one late afternoon as the sun started
to hint at setting. Once majestic, the mansion was now abandoned.
The house's stones had lost their color; they had none of the blues,
grays, or browns that had become so vivid at Isber's. The shutters were
a faded green, the *darabzin* rusted, though somehow intact. The door
had no lock. The house spoke of time's passing, of neglect and humiliation. There was no sound inside save the noise that poured in,
unmediated, from the street. My footsteps were the only ones to have
trod the dust gathered over the years. The only gesture to its former
occupant, a member of parliament until his death, in the 1990s, was
graffiti scrawled on the wall, seemingly long ago.

Here Sat the Tyrant, it read.

As I left the house, my old landlord, Michel Fardisi, approached me, leaving his family's shop across the street. I suspected he was coming to remind me about my last month's rent, which I still owed him. The reminder was always a pantomime. He never asked for money, just rubbed his fingers together as he passed in his Mercedes, his arm extended out the window.

"This house was once filled with people," he said as we stood on the moist gravel of the roadside, near a denuded fig tree that was a few feet from the entrance of the mansion. "They were like ants, like dust settling on the floor." As late as the 1960s, people still gathered here. The house was especially festive in the weeks before an election, when parties rollicking with drink, song, and *dabke* tumbled hours past midnight. "Those days were beautiful," he said. In the intervening years, he understood, twenty-five relatives had inherited it, and no one was close to agreement on what to do with the place, whether to sell it, rebuild it, or leave it to finally fall. "The problems of inheritance," Michel said. "Who's going to fix it?"

When Michel was young, everyone in the area traveled to Marjayoun for most things, old habits not yet broken. It didn't matter whether it was light fixtures or kitchen appliances, jewelry or car parts, even ice cream. Though it had long before begun its slow decline, Marjayoun remained the nexus of the province, and that made it a destination. Michel began to reminisce about those days when not a house in Marjayoun was empty, when bakeries ran out of bread by lunchtime, and when the stores around the Saha were filled — seven tailors, four or five shoe shops, and more than a dozen barbers. In those days, the priest would visit every family, involved in their everyday struggles. Now, he said, the priest sits in his apartment over the church, divorced from the community. The region's nub had become Nabatiyeh in one direction, on the road to Beirut, and Hasbaya in another, on a road that more or less led nowhere.

Marjayoun had been displaced by borders, reduced by time. Most humiliating, perhaps, was the ascent of places past the Boulevard, like Kfar Killa and Qlayaa, forgettable villages that once boasted dirt roads to Marjayoun's paved streets. Now no one asked directions to Marjayoun. *Go past the Saha. After the cemetery. Before the Boulevard. Turn at the Samara house.*

"It's a shame, Marjayoun," Michel said.

Do you like it here? I asked him.

"It's my town. How could I not love it? But I have to work here. If I didn't, I'd leave. What, I'd pull money out from behind the wall? But my son, he won't work here. There's no way. Will he come back? No. The majority of the town is like that." People close their doors, lock them, and leave. "Enough, *khalas,* let's go to America. When the men die, who's going to come and replace them? Not my son, not your sons. After Dr. Khairalla dies — do you know him? He's getting old — will his children come and live here? Not one of them."

Going to America. The words, spoken reluctantly, struck me. Marjayoun's relationship with its diaspora was conflicted, and I heard a hint of bitterness in Michel's voice. Marjayounis took great pride in what their descendants had done — the wealth of the Houranis in the Persian Gulf, and of nearly every family in Kansas and Oklahoma, the Shadids, Naifehs, Samaras, Farhas, Homseys, Rahals, and Shambours. Everyone knew of the exploits of favorite sons, like the late Dr. Michael DeBakey, a world-renowned heart surgeon. Brazil had its Marjayounis, as prosperous as they were numerous. Yet there was a hint of resentment over their abandonment of their homes. No one came back to Marjayoun. When its expatriates did return, the town put on a generous welcome, replete with food and festivities. In Arab fashion, the receptions were lavish. "There was perhaps a little pride, in them and us," Michel told me. There was hope, too, that the émigrés' generosity in return might assist Marjayoun. When one of those expatriates beheld the feast he was offered, he shook his head in amazement. Never had he seen such a spread in America. "I had no idea they were so well off!"

Impressed by their circumstances, he gave nothing to the town.

A few did return and restore their houses. They might spend twenty days here each summer, when Marjayoun, by its more and more modest standards, flourished.

But then what? Michel wondered.

"In the winter, they're going to come?" he asked. "He'll visit his house, he'll stop in, but will he live here? Will he even come every summer? He repaired his house fine, but Jedeida needs people, not renovated houses. Where are the people? We need someone who can come and settle here."

I had returned and rescued a home, in a gesture to history and memory, in the name of an ideal, however misunderstood. But in time I would abandon it, leaving a relic, however functional or beautiful. Each time that happened, this community faded.

Although it had crumbled, someone had resided at Isber's. However unsavory they were, the squatters were part of a community. I would stay in the house, but I would never live in it, and I would never belong.

Michel nodded, as if reading what I was thinking.

"*Hajar bala bashar,*" he said. Stones without people.

Years cannot undo centuries. It takes generations to weaken what has been.

Orthodox rather than Catholic in faith, Marjayoun was always more Anglo-American than French, in education and outlook. American missionaries were influential, founding a Protestant school in 1867. Deemed heretical by the conservative Orthodox clergy, the institution did leave a legacy, the English language — as the Jesuits left an educated class, literate in French, in Beirut and the region known as Mount Lebanon. One of the American missionaries was, for many years, described with the sort of awe reserved for a saint. George E. Post, a Pennsylvanian, was a doctor with a prophet's beard who taught surgery and botany at the American University of Beirut and visited Marjayoun more than once to operate on villagers too poor or too sick to travel.

As the British became ensconced in Palestine after World War I, the town's Anglo-American complexion became, not surprisingly, more pronounced. Ottoman connections and routes shifted to accommodate British borders. Many in Marjayoun remained faithful to the old trails long taken by their forefathers to Palestine and beyond. It was rare to find a family in Marjayoun without at least a peripheral connection to Palestine. Bahija Abla had cousins in Haifa and Bethlehem and a sister in

Beisan, where, after 1948, inhabitants were evicted, fled, or in the case of its Christians, forcibly deported to Nazareth by the nascent Israeli state. With its inhabitants went its name. Beisan became Beit She'an, and after Israel was created, the axis of Marjayoun and Palestine was forever broken, ending an era spanning generations, in which the frontier was never closed by borders.

No one disputed the town's decline. Palestine was gone, along with the lands that Marjayoun's families owned in the Hula Valley and the opportunities its hinterland provided in Haifa, Jerusalem, and the Galilee. There were no more jobs with the Iraq Petroleum Company, no teaching posts to fill. Families returned, or ventured onward — to America and Brazil. In 1967, Marjayoun's other surroundings, the Houran, where Isber Samara roamed, was forever severed, too, when Israel captured the Golan Heights and Quneitra, turning a crossroads into a no man's land. There was no road there anymore, a passage becoming a barrier. The daily taxis stopped. So did the merchants of fortune. The civil war had yet to begin, though, its onset confined to the chatter in cafés, laced with dark auguries and predictions of gloom. Still to come were its horrors and injustices, crimes as vile as any that had been committed in the modern Middle East, names and dates reverberating: 1982, Sabra and Shatila, Tel al-Zaatar.

Weeks had passed since I had last seen my cousin Karim, who was spending most of his time in Beirut. Unfailingly, he would make sure to check in with me, calling at night from a pay phone when the rates were cheaper. "I'm on the corniche doing my fast-paced walking!" he breathlessly declared in one phone call. Again he begged me to water his plants in Marjayoun. "Is that asking too much?" he insisted, already knowing the answer. "Just because you like plants so much," he said.

As summer set in, he finally made his way back to Marjayoun, and I waited for him in his small but lovely garden, amid lemon and orange trees, shapely olives, and pomegranates bearing round red fruit that hung like beautiful jewelry.

Karim greeted me with a requisite complaint about my dearth of calls, but he good-humoredly offered me welcome deliverance. "I know, I know. If I had a cell phone, things would be different," he assured me. "I am nothing if not flexible."

He looked me up and down, narrowing his eyes.

"Believe it or not," he said, pursing his lips, "I missed you!"

He gave me three kisses, a big hug, and asked if I had gained weight. "I missed you!" he said again, then hummed in pleasure.

As usual, Karim brought me gifts: crushed mint that he prepared himself, an empty jar with a yellow lid ("to provide color in your kitchen"), mosquito repellent, and two light green bowls in the shape of either a pear or an eggplant. "For hummus and such stuff," he said. I, in turn, brought him a bag packed with health products — Lubriderm lotion, Centrum Silver vitamins, and a supplement for stiff joints.

The conversations that followed our gift-giving were random but fast and furious, undertaken as we walked around the garden. He pointed to his olive trees, some of whose branches had snapped under the weight of forgotten snows. He was very proud of his clementine tree, which was laden with little white buds. "Wait till you see the pomegranate," he bragged. Then, as was often the case with Karim, the conversation turned to politics. He predicted a war: Israel and the United States on one side, Syria and Iran on the other.

"Ahh, a third world war!" he blurted out.

One day Karim came over to the house and marveled at what we had done, and I realized he really meant it. He seemed stunned. My Lebanese origins were emerging, he declared. "The roots are coming out gradually," Karim said. "You think this is American? This is Lebanese." He pursed his lips again, his head swiveling as he surveyed every corner. "From the bottom of my heart, I'm so happy for you. And I'm proud of you."

He wasn't so happy, though, as to forgo criticisms.

Why did I have a patch of gray cement on the wall outside the house? Couldn't I have found nicer lamps for that room? What about spreading some soil around the garden? Why couldn't I shine the tile a bit more in the salon? "Oh, please, Anthony, you have to. You have to!"

And so on, until he finished with "Really, really I'm so proud of you. You have taste." And I realized that his criticisms weren't suggestions; they were pleas to me to see that he had taste, too.

We sat and had a drink from a bottle of homemade strawberry wine that Dr. Khairalla had given me. We toasted.

Marjayoun was a different town with Karim in it, and the temperature seemed to rise a few degrees with his arrival. He made the air combustible. Despite his origins, Karim never really fit here. However much he wallowed in the age-old rivalries between Hawarna and *baladiya,* the truth is that the town was too small for him. He was too educated, too generous in his own way. A sybarite, he had seen too much of the world. (Or, as Karim once put it to me: "I've ridden on a Greyhound bus, in a plane, on a train, and I've been to an Indian reservation.") Marjayoun never rose to his expectations. Nor did its inhabitants, who were less educated and sophisticated than he, and never respectful enough.

His loneliness and iconoclasm drew me to Karim, as did his enthusiasm. His politics didn't. He seemed to enjoy needling me by stating the most provocative position possible.

"I can't stand the Shia," he said. "I can't stand Hezbollah. I can't stand Nabih Berri. I can't stand Hassan Nasrallah." Even those Shia who claim they are moderate, he went on, are closet Hezbollah supporters. He turned silent, casting a glance at the next table.

"Who is this guy?" he asked. "He could be a Shiite."

"Karim, don't you worry about Lebanon?" I asked.

"Not really," he answered. "I feel there's a truce."

Karim's chauvinism seemed to vacillate. At times he was pessimistic, disenchanted with everything going on in the country, speaking in worried and disillusioned tones. "I don't like the Lebanese people anymore. No, really, I'm fed up with them," he had told me another time. "I'm fed up. We're fed up. I belong to the great majority who are overpowered by the situation." Not on this day, though. It was as if he had thrown down the gauntlet: If no one else would accept him (someone who really never belonged in Lebanon in the first place), then he would accept no one else, and that included Shia, all Muslims, even the Hawarna families in Marjayoun who plotted against the *baladiya.* I felt as though the life he had lived — as an adult, through the country's war and peace — had left him unhappy; politics simply seemed a metaphor for his disenchantment. Like many Greek Orthodox of his generation, he had been an Arab nationalist in his youth. And now "I'm Lebanese first and nothing else," he told me.

But what of his Marjayoun? "I feel like the place is dying," I said.

"No, no, don't say that," he begged me, his voice rising. He still seemed confident, or needed to believe, that Marjayoun would survive, that someday, down the road, it would reassume its place as a crossroads. With a past once so great and so blessed, the town *deserved* to survive. Or so he believed. Marjayoun, he continued, was a meeting place for Palestine, Syria, and Jordan. Marjayounis worked in Haifa, Quneitra, Damascus, and Beirut. In the 1920s, he said, Marjayoun was bigger and more important than Hasbaya and Nabatiyeh. It will be that again, he promised.

I shook my head. "We've been waiting sixty years for that," I told him.

"It's still my hometown," he said. "My roots, my origin, my identity. In Beirut, I'm not Beiruti.

"And this is where I will come when I die. I have a big, beautiful grave ready."

It was never about what Marjayoun was becoming, or whether there would be peace; it was about what it was — a place of parties, meals, guests, and lunches for forty people on Sundays, where everyone seemed to laugh. Karim had his own Jedeida. And whatever I might say about the town's demise, he would never let go of that memory. To hell with the people, he seemed to say; he didn't particularly like them. Yet he loved his Marjayoun.

"What about the town after you die?" I asked him.

"Why should I worry?" He laughed. "When I die, to hell with everyone!"

As the years passed, Abdullah gathered memories that were finally good, the kind worth remembering. His peripatetic existence had come to an end, and he settled in for days that were as relaxed as any he had known since coming to America. He never surrendered his temper, but he spent as much time singing, as friends and relatives gathered at the house over bitter Arabic coffee and bushels of miqta *salted just right. He always seemed to have a cigarette in his hand. With his children gathered around him, he offered each a penny for every gray hair of his they would pluck. He laughed into the night, regaling guests with his voice, sweet and deep.*

Abdullah was never as industrious as his wife, Raeefa. He was as festive as she was serious. He was as temperamental as she was methodical. He would brag about his wife to guests, embarrassing her. Like her father, she preferred silence; too many words were dangerous. The portrait should be posed, revealing only so much. "If you have something good, you don't talk about it," she would say. "Other people will talk about it, but you don't have to talk. Don't say anything." Hearing this, Abdullah would wave his hand dismissively. When she worked in the grocery store, he milled about outside, tending the orchard that was his pride and the garden that was his joy. The grapevines prospered. Though the fruit was never sweet, the vines produced leaves that Raeefa stuffed and then served as a delicacy. Sometimes when he smelled the blossoms — cherries, plums, apples, peaches, and apricots — he told his children that he imagined himself again on a Marjayoun hillside, facing Mount Hermon and feeling the breeze on his face.

20

WORSE TIMES

I HAD JUST A few weeks left to finish the house. And Abu Jean was in charge.

The task at hand was the stairway to the entrance. It was not uncomplicated. Two sidewalks, running perpendicular and at different heights, would join at the stairs, which would ascend three levels. Circular, each would connect the sidewalks at a forty-five-degree angle. Each was of a different length, and my cousins, an engineer and an architect, had carefully sketched the plans on paper.

Abu Jean had no time for all this. (With feigned concentration, he at first looked at the diagram upside down.) He would do it his own way, drawing up the plan with chalk, which soon washed away. When he began, weeks after he had promised to do so, each day seeming to bring another false start, he gave no appearance of being daunted.

None.

He measured the stair. It was supposed to be 220 centimeters long. It was 217.

"*Mazbut!*" he shouted. Correct!

Then he set the spirit level on the concrete that he had poured, to determine if it was level. The bubble in the liquid migrated to the far left. The surface was horribly slanted.

"Good!" he cried with pride, more than a little misplaced.

Fadi showed up to polish the tile we had laid in the salon, dining room, kitchen, and bedroom, but on the third day he declared that he could go no further unless I brought in enough electricity to power his industrial equipment. Busy, I asked him to go to the utility company, *please*. He deferred. He wasn't from the town, *ibn Marjayoun,* he said. Abu Jean didn't want to go. The house was in my name, after all, he pointed out. So I went, and ten minutes later we had power. Wonderful, Fadi said. But he still wanted to wait until Malik finished laying every piece of tile. Malik shook his head: He was waiting for two doorstops from the neighboring town of Khiam. And by the way, he told me, he needed more sponges. Not surprisingly, Abu Jean was reluctant to proceed with the stairs until Malik finished tiling the sidewalk. Malik wouldn't start tiling the sidewalk until Abu Jean poured the concrete. After all this time, neither knew what the stairs were supposed to look like.

A sane man would have admitted the obvious: I couldn't finish before I had to leave Marjayoun, my sojourn over. But sanity wasn't figuring in my thinking. We can finish, I kept telling myself, and the only person more insufferable than Abu Jean was me. The phones were abysmal in southern Lebanon; the network was neglected by the phone company and (supposedly) jammed by the Israeli army and United Nations peacekeepers. I tried anyway, making scores of calls on a phone missing its battery cover. I called Marwan to find out when he would take the doors to his workshop and paint them. I called Toama's cousin to check whether he had the right paint colors for the walls — soft rose, stone, and cream. I called — *begged* — Nassib to bring the restored *darabzin* for the balconies and windows. Every day I asked Ramzi when he would finish the wood ceiling. He humored me, repeatedly telling me he would meet a deadline and then always missing it. Camille the carpenter, his delays epic, defying the odds — what were the chances that *every* deadline would be missed? — told me he couldn't come because his newborn cousin had died.

"*Allah yirhamu,*" I said. God have mercy on him. Then, without pausing, I asked if he could send his Egyptian apprentice, Shawki, instead. We had a lot of work to do.

I exhorted the company in Beirut to finish the windows, shutters, and wooden tracery of the arches. I begged a company in Sidon to put the parquet in another bedroom. They did, only to complain that Abu Jean's cement floor was too shoddy. Abu Jean, in turn, erupted in a fit of self-righteousness, running his hand over the floor's wadis in admiration.

"It's smooth as silk," he insisted.

Then my Shadid side ruptured.

Toama and Abu Jean were fighting once again, as usual over money. Abu Jean was zealous in not spending, as he put it, a franc more than he had to. Toama was no less stringent in charging me for any task, no matter how trivial. He even charged me for gas to drive a few minutes away to check whether the stone molding was ready. On this day, they argued about cement. Toama wanted Abu Jean to buy another bag. Abu Jean insisted there was enough left upstairs. *We need more,* Toama said. *We have enough,* Abu Jean said. The fight went on. Neither said anything different; only their voices rose. On and on, minute after minute.

"Enough!" I shouted.

I took my wallet out of my pocket and threw it to the ground.

"Take the money," I said.

My show of anger made Toama lose his temper. He walked away, shouting. So did Abu Jean. With four words and a dirty billfold, I had inaugurated a procession of hurt feelings, wounded pride, and bruised egos that ended the day's work. Others tried to mediate. Appeals were delivered to God. Coffee was brewed. To no avail.

Then George Jaradi drove up in his derelict white Mercedes, which he had named for a prostitute whom he seemed especially fond of. Jameela, he called it.

"How are we?" he asked.

Even in the most despondent times, George made me smile. I patted his ever-growing gut. "When's the due date?" I asked him, drawing a cry of glee.

Briefed by all, George would be mediator.

"What has George told you every day, Anthony? George is working here in Marjayoun, at a *warshe* up there," he said, pointing toward the Boulevard. "When there's a problem, call George, and George will fix it for you. Didn't George tell you that? Don't get angry, don't let your head hurt, Anthony."

We sat under the olive tree, its branches laden with fruit.

"George wants to talk straight. Neither left, neither right, but straight. George will come and finish it for you without pay. Just call George," he said. "George will do it."

I knew he wouldn't, but I appreciated the offer. Like my grandfather, I cooled as quickly as I got angry. We talked a little longer, and I promised I wasn't mad at anyone.

"The yogurt's pure?" he asked in Arabic.

It meant there were no more grudges.

I nodded my head: The yogurt is pure.

When George left, Abu Jean muttered into my ear, "Fuck him!"

But the next day, neither Toama nor I nor Abu Jean mentioned a word about the bag of cement.

It was May when I visited Dr. Khairalla. I had made a CD of songs by Rahim Alhaj, the virtuoso Iraqi oud player, to give him as a gesture of thanks. When I asked him how he was feeling, his first urge was to tell me he felt fine, but more than a suggestion of doubt passed his face, and he spoke more honestly.

"Not so good."

The cancer had spread to his lower spine (or, as he put it, L4, the fourth lumbar vertebra). The pain was excruciating, and he seemed changed. His thick gray hair was still combed back from each side of the part, like an actor from the silent-film era or a dapper socialite in 1920s New York. But his face was darker, more drawn, and I could see he was weaker. He hesitated in his step, and his reactions, the quick movements I remembered as he pruned the plum trees at my house, were more subdued.

He was sitting with Ali, a Shiite friend of the family, from a neighboring village, who had brought his son for a medical exam. Ali turned to me. "There is no one like him," he said, nodding toward Dr. Khairalla. As they left, without a suggestion of payment, Dr. Khairalla slapped the boy on his butt, the first time I had seen him being playful or energetic since we had met.

Ivanka served coffee and brought out a plate of deep red plums that tasted like sugar. Dr. Khairalla showed me a picture of the bouzouki he had made for his grandson, Jean, who would turn four this year. Crafted of wild plum wood, the handle was dark. The face was made of a light, more fragile wood. The base was a gourd. "From Grandpa to Jean," he had inscribed on it in Arabic.

My phone rang, with another message about the crisis in Lebanon, the stalemate having given way to a new bout of tension. There were rumors that a strike the opposition had called might turn violent in the capital. Dr. Khairalla had heard the same, and he was bitter about it.

"We've become the stage of the theater where they act," he said.

"Will there be a civil war, Dr. Khairalla?"

"It's not in the hands of the Lebanese. The decision is outside," he said. "We're the instruments of war, the means. They use us."

For weeks I had asked Dr. Khairalla if I could see his bonsai collection, which he had begun in 1990. Finally he broke down and gave me a look, beginning with the cotoneaster he had bought with me in Jibchit. He had already pruned it and was beginning to train it, wrapping aluminum wire around certain branches to give it the shape he wanted. He went from plant to plant, describing them as if for a catalogue, much the way he did the garden. There was a satisfaction in his words, a sense of achievement; he talked about the small plants the way he talked about the ouds he had built. He had a two-year-old wisteria he was trying to train. There was an acacia, a fig, and an apricot tree, a jasmine and a rose without thorns. "Look," he said when I seemed distracted by phone messages about the crisis. He pointed at a wild plum, and a magnificent olive tree no bigger than a beer mug, and the oldest of his plants, a ten-year-old cherry tree. The way he talked about them all, it seemed he still saw himself as a beginner. A real bonsai, he said, would take at least ten years to perfect.

Ten years, I thought. He had just started on the cotoneaster a couple of weeks ago.

"There are bonsai one hundred, one hundred fifty years old," he said. "You have to be patient. This is the principle." When he told other people about his collection of thirty plants, they dismissed the idea of spending so long on something. "They want everything quickly," he told me. He understood the investment; he had spent two or three years just deciding whether to begin in the first place. "I was afraid of spoiling the plant," he said.

We stood in silence, staring at their delicate precision.

"It is a kind of meditation," he said, then laughed. "I deal with plants like living creatures. I feel pity for each one. Especially if you grow it yourself and it reaches a certain age and it dies, you feel like you're losing something that interacts with you. It's not a nice feeling to have."

My phone rang once more. Another world intruded again, and I had to leave.

I could tell he was growing tired. He said his doctor had urged him to take morphine, and he had refused. "It's not wise to take morphine now," he told me a little clinically. A doctor attending to his own condition, he knew his prognosis. "If I take morphine now, what will I take later? I'll need something else." He paused as if thinking about what lay ahead.

"There will come a worse time."

The picture reads Wingfield's Studio, Oklahoma City. *The inscription is written in cursive in the bottom right-hand corner. The portrait features Raeefa, in her thirties now. She remains attractive, though she has lost some of her youthful features. Her hair is shorter, styled and swept back. Her face is heavier, as are her shoulders and arms, which were once so delicate. One feature in the more recent picture haunts: her eyes. Shimmering, they have a faraway look, a distance that speaks of hardship. There is an emotion familiar to me from the eyes of people of war, victims of violence and loss, witnesses to simply too much. Still young, five children hers, Raeefa had already seen what she rarely cared to recall.*

It was January 1942, and for the first time since they left Lebanon,

Raeefa and Abdullah had settled in Oklahoma City, which was booming thanks to the stockyards and the discovery of oil, a share of it underneath the state capitol itself. Their modest contribution was the construction of a building that housed a grocery store. Next door they rented rooms to Doc Roe, a pharmacist who established a drugstore. In back was the orchard, Abdullah's pride. Next to it was the house, modest, simple, and luxurious by the standards of their years of wandering.

Lately Abdullah had been complaining about his health. Just three weeks before, he had gone to the Veterans Hospital in Shawnee, Oklahoma, for a checkup, and some nights he would lie in bed singing lonesome songs in the Arabic of his youth, his own plea to stop the pain. Nothing bothered him on this day in January, however. He sang and he laughed. He joked in the store. He was filled with euphoria, trotting through his orchard, even though the trees' branches were like barren webs, colored gray. He was at ease, conversation effortless, whether it was with Raeefa, the children, or their customers.

Nightfall ended his respite, and he woke up suddenly. He felt as if he had indigestion, and there was a gnawing ache in his chest, the kind that makes you hold your breath, hoping the pain will recede. Things got worse. "I'm hurting," he told Raeefa. "My chest is hurting."

Grimacing, Abdullah got out of bed. He walked slowly, almost a shuffle, to the next room. He sat down on the carpet beside a brown gas stove, then lay on his back. Raeefa followed, staring into the darkness, crouching next to him.

"Come back to bed," she said to him in Arabic.

There was no answer. Not a moment more passed before she knew.

"Abdullah!" she shouted. "Abdullah!"

The children awoke and ran into the room. The youngest son lifted his father's eyelid, having seen the same gesture in movies. Just thirteen, the oldest son tried to resuscitate him, his attempt more desperate than effective. Raeefa rubbed Abdullah's hand, even then losing its warmth. An ambulance eventually made its way to Northwest Tenth Street, but it was too late. At forty-nine, Abdullah Ayyash Shadid, immigrant, soldier, oil worker, grocer, gardener, and wanderer, was dead.

"He was in my arms when he passed away," Raeefa would always remember.

Weeks later, at midnight, Raeefa lay in bed, unable to sleep. One thought after another unfurled — money, her children, a solitary future. She walked outside, leaving the front door open. It must have felt as though the house was suffocating her. Under the stars, her voice could be heard. For fifteen minutes she looked toward a cloudless and moonless winter sky, in an unrequited stare, searching. As fate loomed, her arms were outstretched.

"Allah yisaidnee," she said. "God help me. Guide me. Help me carry on."

Her oldest daughter found her and helped her back inside.

When she woke up the next morning, she said, "I'm going to open the store."

So much of the house was what you might call memories of what I had imagined over many years. I spent hours at a time walking through the garden in Marjayoun. Sometimes I thought back to the lazy evenings of a humid summer in suburban Maryland with Laila, my wife at the time, and my mother sitting idly on the brick porch of our old home. Next to me were the *miqta* that never bore fruit but battled their way through the garden. Amid them were the tomatoes that grew taller than me, their fruit-laden branches propped up by wire, string, and stick, as fireflies flashed in the dark.

By now, there were dozens of *miqta* snaking along the ground in my Marjayoun garden. Tomatoes that Malik gave me were bearing flowers. Thanaya had planted patches with parsley, coriander, onion, and garlic. The oranges — the clementine and the Abu Surra — had white buds, which dared themselves to become flowers. Abu Jean had set down a stone stair at the garden's edge, and as might be predicted, it was askew. As I stepped down on it from the sidewalk, I heard the muezzin in the distance. His voice was lonesome, barely audible. The only hint of the outside world was the U.N. helicopter that rumbled above, its rotors making a sound that was more vibration than noise. I thought of the day, nearly two years before, when I had arrived at Isber's house to find the damage wrought by the solitary rocket. I remembered the sound of helicopter rotors then as I sat on the step and tried to take in the scene.

Between the two olive trees, inherited from Isber and remembered by Raeefa, I set up a table. For a top, I used a slab of marble from the counter of Abu Elie's old kitchen that we had dismantled long ago. I bought three plastic white chairs from a shop down the street. On this evening, I sat there and stared into the distance, not wanting to be anywhere else.

After the death of Abdullah, life for my grandmother continued as it had been in many ways, simple and otherwise. At the store, most of the customers used credit, and after Abdullah's death, many of them paid off their debts, a gesture of sympathy for the widow and her five children. Raeefa's brother Nabeeh came each night to the store, modest even for its day — twenty, maybe twenty-five feet by sixty feet, far smaller than a modern convenience store, but built to last. Like her father building his legacy in Marjayoun, Raeefa let nothing be discarded when it was time for repairs or remodeling. Half bricks were reused and the walls still stand seventy years later. In time, Nabeeh taught Raeefa's oldest son to work as a butcher, and he assumed some of Abdullah's duties, taking a hindquarter or a full side and cutting it into round steaks, T-bones, sirloins, and roasts, which were stored or displayed. On Saturdays a customer brought chickens, usually twenty or thirty. With a knife or by hand, the young man slit or wrung their necks. The second-oldest son worked with the produce. Nothing was wasted: Carrots, radishes, and celery were cut back then rebagged, and lettuce was peeled then resold.

No prices were posted. Ringing up the goods, one of the two sons would shout to Raeefa, "How much is a can of beans?" or "How much is toilet paper?" She would shout back the price; she knew them all. At night the three of them would sit around the black potbellied stove, responsible for scorching the coats of endless customers. They talked back and forth until 8 P.M., or whenever the last customer left. Then they huddled around a radio, listening to Amos 'n' Andy before going to bed. For a time, Raeefa passed a Bible around, and each child took turns reading a verse.

For years they had no car, and Raeefa's only social life was when people came to visit. In the summer many did, sharing the miqta that always seemed abundant. On those hot nights, the air still, the children dragged a wine-red Persian carpet and their bedsheets outside to sleep on the grass. By this time, the Lebanese community in Oklahoma had taken shape. The services that Khoury Shukrallah performed in his house for eleven years soon attracted so many of the devout and devoted that a church had to be built. It was originally envisioned as Saint George's, named after the venerable Orthodox church in Marjayoun, built the century before. But the Greeks of Oklahoma City had already claimed the name, so Saint Elijah's was chosen instead. Dedicated on September 14, 1931, the small frame structure was thirty feet by forty feet. It cost $2,000. The building contained two icons, Saint Elijah and Saint George, the patron saint of the Marjayoun church.

Every community needs a miracle, and Saint Elijah's had its own. In November 1935, rumors spread. The Daily Oklahoman reported that neighbors of the church, Mr. and Mrs. A. S. Bell and Mrs. G. W. Croskery, were awakened late one night by a bell's toll. Sounding for more than a half hour, it kept them awake. They called Khoury Shukrallah. Please, they asked, don't ring the church's bell at midnight. He shook his head, wondering if he had misunderstood their English. Saint Elijah's has no bell, he told them. Two days later, the priest's neighbors heard the bell tolling again, and Khoury Shukrallah offered to give them a tour of the church to prove none existed. Both priest and congregation saw it as a sign from their patron saint. "We are going to build a bell," Khoury Shukrallah told a reporter. "We have received the command." Less than three months later, the church had a steeple and bell. Raeefa would hear it toll on Sundays, when her brother Nabeeh, who served as a father to her children, would drive to Northwest Tenth Street and take her to church.

21

IN THE NAME OF THE FATHER

I WAS IN BEIT MERI for the baptism of Hikmat's daughter, Miana Maria Ruth Farha, at the Mar Elias Church, built amid the Roman and Byzantine ruins of the town and perched over Beirut and the Mediterranean Sea. Sheltered by three olive trees, the church was less remarkable than the view, but still stately, adorned with red Persian carpets and thirteen rows of simple, austere wooden benches. The altar was of hand-carved dark wood with vine and flower designs, arrayed with the icons of Christ, his disciples, and patron saints.

Icons lined the walls of the church, built in 1872. Mar Elias carried a sword in a martial pose. Saint George, the Christian legionary from Roman Palestine who was martyred for refusing to worship the old pagan gods, faced the ubiquitous dragon. Four censers hung low, and a fifth dangled from a cross, bound by three chains meant to symbolize the Trinity. From the time of the Apostles, incense was burned, and here it wafted through the nave. It symbolized prayers lifted to God and the grace of the Holy Spirit embracing them as they ascended.

The smell of the incense took me back to childhood, when once I peered at row after row of candles at the entrance of Saint Elijah's in

Oklahoma. Even there, the rhythm of the chants was Eastern, recalling a distant time and lost home, with its suggestions of Byzantium and Constantinople, and its inflections of Greece, Rome, Persia, and the ebb and flow of Arab tribes moving across the Syrian desert. For a moment, there in Beit Meri, I heard what felt like the echoes of chants sung during Ashura, the holiest time in the Shiite Muslim calendar, as tens of thousands of Iraqis marched under green, black, and red banners, beating their chests as they surged into a shrine of gold-leafed domes and minarets. I was in Karbala again, covering the war, wandering through the mournful air of those laments, staring at flags that bore saints' names, blood falling from the letters in a symbol of their martyrdom.

Hikmat was dressed in a gray suit with a black and gray tie lined with pink. Amina was in a black, loose-fitting dress. Miana wore a white bib, and as the ceremony began, her expression suggested that she actually understood everything that was transpiring. Like my daughter, she seemed older than the calendar suggested. Father Haris, with a black beard, wore a gold sash with a purple border. The other priest, Father Philippos, was older. "I love him," Hikmat told me. "You know why? He used to come sit with my father."

The ceremony was elaborate, as most Orthodox rituals are. The women of Hikmat's clan were gathered around the baby. Miana was one of theirs, and I could feel the bond between the women and the lovely child. As the baby cried, Father Philippos joined them. His multicolored cross hung low on his chest, his purple robe bordered in gold. He started singing, in a voice that was beautiful and melodic, with a softness that suggested the tolerance of great age.

For a minute, miraculously, Miana stopped crying.

Father Haris rolled up his sleeves, beginning the sacrament of Holy Illumination, little changed in a millennium and a half, to transform the old and sinful into the new and pure. The priest breathed three times on Miana, then made the sign of the cross over her.

In the name of the Father . . .

The day was radiant, as it often is in spring. Weeks before the baptism, Hikmat and I had sat on his porch in the morning, sharing tea. Judging the weather too mild, Hikmat put a thimbleful of scotch in both our mugs. We talked about respect for others, and I realized he had thought a great deal about it. "You have to give respect," he told me. He offered his code, which I suspected was inherited: Be respectful, don't be aggressive, help people through your support. "Show them care, show them care," he said. "And one day, they will return your respect." He mentioned an Arabic proverb that I had heard him pronounce often: "He will dig up your grandfather's tomb." If you cross someone, he won't relent until he finds the ransom of your family's secrets. With those secrets, in a town too small, he can ruin you.

"In the end, we're judged by society," he said. "Society is God.

"The Farhas in Marjayoun, the exceptional house is ours. *Intu Farha moumayazeen.* My father and his family are beloved," he said. "My father was exceptional." He caught himself, as though he might be bragging. Ask George Abla, he said, or Abu Jean. "I see other good fathers, everyone has a father, but no one is better than my father."

Father Haris placed his right hand on Miana's head.

In your name, Lord God of truth, and in the name of your only begotten Son and of your Holy Spirit, I lay my hand on this your handmaid, Miana, who has been found worthy to appeal to your holy name and to seek shelter in the shadow of your wings.

Hikmat's father, George Mitri Farha, was born in Marjayoun in 1906, the second oldest in a family of seven. Of the siblings, only George had children. "We were raised by all of his brothers and sisters," Hikmat recalled. "In those days, wealth was food, not money. Property and food." George Farha named his oldest child Rifaat. Hikmat followed — like his father, the secondborn. "You learn love from your family," Hikmat told me. "My father loved us. No one dared lay a finger on us because they knew my father would kill them."

Handsome and taller than Hikmat, George Farha presided over the good years in Marjayoun before the civil war. Back then, Beirut was five hours away by car, giving Marjayoun a greater notion of independence. It took care of itself. As Hikmat's mother recalled, there were *shakhsiyyat* and *ailat* back then, personalities and families who lived large. Everyone went to weddings, and no one hesitated to attend funerals. Across the street from George Farha's house, Michel Shambour, almost deaf, listened to his radio. His ear hovered near the speaker, turned as loud as it could go. The town was filled with noise.

"We lived good in Marjayoun," Hikmat said.

During the civil war, the bad years that stood as an epitaph to Marjayoun's long demise, George Farha helped protect residents from kidnapping by militias. Without bias, he extended aid to Sunni, Shia, Orthodox, and Maronite alike. "My father used to support the weak and the right against the wrong," Hikmat said. "People used to respect his word. He was a trustworthy man."

When the civil war started, Hikmat wanted to fight. Impetuous, he went looking for weapons, but his father stopped him. "When two countries fight, hide your head," he told him. "*Ya ibnee,* this is not our war."

In 1976, the family fled the war for Beit Meri. Every night before his father went to sleep, he prayed with his eyes closed, then made a cross in the air for each of the children who had left the house. "Hala in Dubai, Hikmat in Barbados, Rifaat in Switzerland," he recited. "I believe I'm still living because of his prayers," Hikmat told me.

George Farha died in 1993 of pneumonia.

"Men were different back then, no?" Hikmat said.

Father Haris began the first of four prayers. Miana began crying, and Hikmat walked away, as if he could not bear to see his daughter in discomfort. The priest spoke again.

The Lord condemns you, Satan.

Hikmat had another role model: his maternal grandfather, from the Obeid family in Wadi Nasara, a predominantly Christian region of rolling hills in western Syria, home to the Crusader-era fortress Krak des Chevaliers. By Hikmat's account, Hana Obeid was tough, a tall, big man with a grandiose handlebar mustache who never smiled. He had asthma and, Hikmat recalled, "he breathed like a lion." He had a gang, too, a bunch of menacing ruffians. As the story goes, when a Greek Orthodox bishop in the region insulted the Obeid family, Hana decided to send him a message: "Behave yourself, respect yourself."

His plea went unheeded, and the insults continued.

"My grandfather called his men. They went to Deir Mar Giryus, an ancient monastery with two churches, one eight hundred years old. You have two roads from the *deir*. He told his men, go and block the roads and teach the bishop a lesson. Hit him. Break him down."

The men went, a dozen or so on each road. Eventually the bishop left the monastery, but was stopped by piles of stones blocking the road. Hana's men grabbed the bishop's driver and put a pistol to his head. "Don't move, you son of a bitch," one of them said. Then they shouted at the bishop, "Come down, you devil's beard. Come down."

The bishop did, bowing on his knees. "I swear by Hana Obeid, don't hurt me," he pleaded with Hana's men. Hikmat continued: "One of those men. What was his job? Barber." Hikmat slapped my knee with an open hand. "He takes the scissors out of his pocket and he cuts the bishop's beard to make him look like a devil. This was in the days of the French. Nobody dared to do it. He cut the priest's beard!"

Father Haris again blew on Miana three times.

Expel from her every evil and unclean spirit hiding and lurking in her heart, he said each time, Miana's eyes transfixed on the priest's vigorous black beard.

The spirit of falsehood, the spirit of guile, the spirit of idolatry and greed, the spirit of deceit and all impurity prompted by the Evil One.

Endow her with reason, as a sheep of your holy flock, an honorable member of your church, a dedicated vessel, a child of light and heir to your kingdom.

A picture of Hikmat's father hung in his house in Marjayoun, to the left of the fireplace. A rosary hung from the portrait's right corner. Hikmat and his father shared the Farha nose.

"Marjayoun is my father's house," Hikmat once told me. "Marjayoun is my father."

Hikmat fostered his father's legacy as the *zaim* of the town. Hikmat would *be* his father, inheriting the respect of the idyllic Marjayoun of his youth. He rebuilt the house, he won a seat on the town council, he spoke in his father's name.

"Marjayoun," he said again, "is my father."

Satan was renounced, and Christ was joined. Father Haris called on all who entered with "faith, reverence, and godly fear" to join him in praying to the Lord.

That she, and we, may be spared from all affliction, wrath, danger, and want.

"I accept what God gave me," Hikmat told me. "Do you understand what I'm saying, *khayee?* Do you agree with me?"

He talked about what God gave us, what we can rely on. To Hikmat, it was family, pride, reputation, and dignity. Family, he said, came first, the source of the others.

"A tree without leaves is naked," he said.

I thought of this as I watched him in the church, crossing himself on the chest.

Indefatigable, Father Haris pressed on as the baptismal water was blessed.

And give to it the grace of redemption, the blessing of Jordan. Make it a fountain of immortality, a gift of sanctification, for the remission of sins, protection against infirmities, destructive to evil forces, inaccessible to opposing power, filled with angelic might.

Father Haris made the sign of the cross over the olive oil, then poured it in the shape of a cross into the baptismal water three times before taking Miana in his hands.

Hikmat and I often talked about what might save Marjayoun. I was never all that hopeful. Like Cecil, Hikmat was. We talked about a windmill farm, exploiting the gusts that barreled through the Litani Valley. He once had an idea of fashioning blocks of wood from the remains of pressed olives, the *jift* that was plentiful in the region. For years he had wished for a college, even a university, in the town.

"Our values matter, and we can't lose them," Hikmat told me over drinks one day. "Marjayoun has lost almost everything, but it still has those values." Like others in the town, he credited the Bedouin culture that the Hawarna brought from the faraway steppe in Syria. "As Christians, we are basically Bedouins. We know the Muslims. Not only that, we behave like them." The traditions distinguished Marjayoun and he praised what they represented.

"You cannot ask for a favor or a service unless you spend three nights at someone's house. You have to eat with them, sleep with them, then you can ask.

"These cultural things, time will not erase them easily. Did I tell you our house in Marjayoun is older than America? Four hundred years. It might sound silly, but I'm proud of it. Get help and give help. Human values, not money values, technological values, machine values. These things are worth something. This culture matters to us." I poured him another drink as the hour grew late. "Hopefully, they won't go," he said.

Father Haris submerged a crying Miana three times in the water. She stopped crying, and the women of Bayt Farha cast a look of horror; several thought she had stopped breathing. The priest reassured them. "No one has ever died from a baptism," he said. Miana started crying again, but less so, still shaken.

He anointed her with oil and she was soon dressed in her baptismal gown.

The Lord is my light and my salvation, went the chant as the ceremony approached its end. *Whom shall I fear? The Lord is the protector of my life. Of whom shall I be afraid?*

One day in the late afternoon we were sitting on Hikmat's porch, snacking on stuffed grape leaves and pastries filled with thyme and cheese. Fahima, who lived upstairs from Hikmat, brought salt-cured black olives that she had prepared.

As she set them out in a bowl behind him, Hikmat turned toward her.

"Was my father good?" he asked.

Tears welled up in her eyes.

"You'll never be your father," Fahima told him. There was no cruelty, no venom in her words. "You're a good man, but you'll never be the man your father was."

"Life changes," Hikmat said, not disagreeing. "I'm fifty-five years old. Forty years ago, I would say a word and I'd close the market. If somebody did this to me," he said, gesturing as though he was being slapped, "I'd kill him or I'd die. Not only me. My brothers, too. What are these men here?"

Soon after the baptism, I noticed an envelope pasted to my door addressed to Najib and Nabeeh Samara, my grandmother's two brothers. It was a water bill, listing the charges from 2002 to 2008, a total of $1,030. I joked with Shibil that I should have written across the top, *Ahlan wa Sahlan,* Welcome to Marjayoun.

Shibil was blunter. "It's a fuck-you for coming to fix your house."

In matters like this, I always needed help, and I went to ask Hikmat if he would join me at the water company, near my old apartment. Surprisingly, he hesitated for a moment, then agreed. He had never been there, and a little meekly we walked around the old villa that housed it until we found the entrance. A knot of employees sat outside, drinking bitter coffee under the sun, near a sign that read, *Don't enter unless you have business here.*

Sounding gracious and warm, Hikmat sought a person to talk to. Most stared at their coffee cups, and some looked away. One of them tossed his hand, pointing him inside. There, a man, made lazy by boredom, was sitting behind a rickety desk.

Hikmat said the right things. Why should he pay when his house was wrecked and abandoned? Why should he bear responsibility for his ancestors? Why should he pay a bill that was not in his name? Younger than Hikmat, the bureaucrat sat behind the desk, shaking his head in the passive-aggressiveness so familiar in any government office. *To provide help is to lose power.*

Every transaction in these parts requires a bit of subtlety. After finding out the man was from Taibe, which meant he was Shiite, Hikmat suspected that the bureaucrat's politics mirrored his sect. He was almost certainly a supporter of Hezbollah, he guessed, probably putting him in opposition to the government. Nodding, Hikmat started criticizing the state, all the thieving ministers and, most of all, that wretched prime minister, Fouad Siniora. All the while, he hoped he had read his interlocutor right. Torpid, nearly lifeless, the unsmiling man simply nodded.

Weary of the conversation, the employee finally wrote a letter that we needed to have signed by the mayor. He weighed each word he wrote; he squinted after finishing each sentence; he spent fifteen minutes on a half-dozen lines of text, rendered in tiny, precise handwriting. Hikmat could leave without having been humiliated, or could claim as much. But I knew as well as he did that the paper was just that, paper. I would have to pay.

We sat at his house afterward, drinking bitter Turkish coffee.

"A small piece of shit in the government, and I can't control him?" Hikmat asked angrily. "You think if it was in the days of my father we'd have to pay a thousand dollars?"

"What would your father have done, Hikmat?"

"My father would have gone about it a different way. He wouldn't have paid. He would have contacted a minister. He wouldn't talk," Hikmat shook his hand, "*haik*." It meant like so. It meant the way Hikmat had talked.

We sat at his white plastic table on the balcony where we had spent so many hours. The weather was still cool, bringing a sharp breeze to his fifteen dunums of land spilling out in stone terraces beneath his house. Hikmat and the gardener had planted cantaloupe, and he was excited that they were already bearing fist-size fruit.

"Are your father's days gone forever, Hikmat?"

He blinked his eyes, a gesture in the affirmative.

"Different men, different people, different world, different society, different mentality. The young used to respect the old. Different everything. What is *akhlaq* in English?" Morals, I said. "Ahh, morals. Now the morals of the people are to be a crook."

I had heard the sentiment often, from Malik, Assaad, Shibil, and now Hikmat, a man who had more faith than they. To him morality manifested itself in the trivial. "If I saw a thirty-year-old and I was eighteen," he said, "I would put my cigarette out in the street.

"We are proud of what our grandparents did, Anthony." They were the equivalent of nobles, he told me, men of stature and influence, whose very presence would change the tenor and tone of a conversation. "This house," he said, pointing next door to the abandoned Farha villa, a palace in its own right, "who can build it from this generation? Each stone of this cost one gold coin. One gold coin! Think of each stone. This is their gold. They left their gold there."

As we talked, the wind picked up, blowing louder through the poplar trees. In a white T-shirt and pressed green slacks, Hikmat lit a cigarette from his pack of Gauloises.

"If I was smoking a cigarette and an old man passed, I would hide it," he told me again. "It's impolite to smoke in front of a man older than you."

He shook his head. There was nothing more for Hikmat to say.

22

COMING HOME

A S WE BARRELED toward the house's completion, the days were full of tumult, the kind that could revive any spirit. The prolonged periods of inactivity just weeks before had terrified me, but now things were different. There was action. Brooms swept; crowbars ripped down scaffoldings. Bags of cement were opened. Stacks of tile were unbound. Molding was unsheathed, and pieces of marble were set into place.

So much was done. A different room was painted each day, though none were yet finished. (The task was made more difficult by a revelation coming rather late: Toama, who was helping paint, was colorblind, which became clear when a room was rendered in three colors.) Tireless and still gruffly intimidating, Malik tiled the rest of the upstairs: salon, kitchen, dining room, bedroom, and balconies. Fadi began cleaning the tile, which would take days and days. At first he simply swept away the dirt, but even then, the change was miraculous. Patterns emerged, in a coy way, flirting and suggestive. They hinted and teased, although their glory still waited to be discovered. Nassib Subhiyya, the blacksmith, began installing the antique *darabzin,* newly painted black, on the balconies — another vestige, one of the few left, of Bahija's day.

A jarring number of tasks remained, with just a few weeks to finish them. There was the mundane — installing the kitchen counter and cabinets — and the exquisite — refitting the arches to match the way they were when Isber Samara built the house. We needed to finish the roof, paint the doors, build a staircase, tile the driveway, fasten the *darabzin* in the stone molding of the windows, add two more coats of paint to the walls, and repair the ornate vents made of gypsum in the salon. Power cords snaked across marble that had yet to be polished. And that metal barrel — the same metal barrel that had occupied the very same spot since February — sat in the *liwan* in all its rusted glory, filled with swampy water. Yet for the first time, perhaps, I could see the very end.

On a day in May, Abu Jean and I drove to Kfar Killa, a few miles away, on a road along the Israeli border that twisted and turned. We had to pick up more tile for Malik, who complained about trying to tile the crooked walls in the kitchen that Abu Jean had built of cinderblock.

"I should spank his ass," Abu Jean said, in words that were a little slurred. For a week, he had come to work without his dentures.

We entered the town of Qlayaa, and he asked me if I knew what its inhabitants once did. I had heard this story countless times, on drive after drive.

"Ah, Abu Jean," I said. "They put their hands behind a cow's ass, let the shit fall in them, and then plastered their walls with it. That's how they built their houses."

Abu Jean clapped and laughed with gusto. It was the happiest I had seen him. In his eyes, I had at last become a Marjayouni: I knew how to insult the neighbors.

We arrived in Kfar Killa. I paid for the tile, and an employee offered us coffee. In a gulp, I drank it and got up to leave. Abu Jean looked me in the eye. That was all he had to do to tell me that he was not ready.

He sat back in the chair as if he was at his mother's house. He drank his coffee as though he was savoring every drop. Between each sip, he drew on his Cedar cigarette as if it was his last. The ash grew ever longer, seeming to burn by a different measure of time.

The leather-covered pocket-size notebook is in remarkable condition. Its cover is wrinkled, as would be expected after fifty years, but it remains otherwise intact. It carries the details of the trip that Raeefa, then fifty-two, her children no longer living in her house, embarked on to Europe and the Middle East.

A decade before, she had closed the grocery store, having earned enough to ensure that her three sons were on their way to becoming a lawyer, a dentist, and a doctor. She was a wealthy woman with a keen eye for real estate and investments. She lived in Oklahoma City's best neighborhood. Forty years before, as a scared girl clinging to her father's gold and her mother's jewelry, Raeefa had traveled by boat across the world. This time she returned on an airplane, her first experience of flight.

Her passport still charts her itinerary for that trip: Great Britain, Switzerland, Italy, and France, which she once visited on her way to an unwelcoming New York. From there, the document records her stops across a Levant whose borders were being drawn when she left: Lebanon, Syria, Jordan, and Egypt. The borders would change again. In 1960, East Jerusalem belonged to the Hashemite Kingdom of Jordan, as did its now closed airport. Egypt and Syria comprised the United Arab Republic; that vision of Gamal Abdel-Nasser would not dissolve for another year.

Scribbled between her diary's pages are the names of people she planned to see. Her Arabic sentences are rendered as hesitantly as her English ones. In Saydnaya, she writes, she would visit al-Hajja Maria, head of the convent. Her friends made certain she would be received wherever she chose to go. Zaki Naifeh jotted a note to a merchant in Beirut, Fouad Nasr, who ran his store at 22 Souq Sursock, in the Beirut neighborhood of Ashrafieh. "My dear Fouad," he wrote, "I'm sending to you our friend Raeefa, the wife of Abdullah Shadid. She is like a dear sister to us. As you treat your brother, you should treat her."

With her sister Nabiha, who left Marjayoun with Nabeeh, Raeefa visited their youngest brother, Najib, in Egypt. He was working for the Iraq Petroleum Company in Syria, but his family lived in Cairo. Raeefa had written their address in her notebook: Asma Samara, 17 Kasr El Ali Street, Garden City, Cairo, U.A.R. Najib had been a child when they left; he was approaching his fifties when they returned. The sisters were sitting in his apartment drinking coffee, the American variety, when he walked through the tall, turn-of-the-century doors. For the first time in his life, at least that his children could recall, he cried.

It was August when Raeefa and her sister took a chauffeur-driven car from Beirut to Marjayoun. When she had left so many years before, Raeefa sat in the horse-drawn buggy, tears shrouding her vision. This time she saw the vistas for which the country was fabled — the turquoise Mediterranean as she approached the port of Sidon; the cliffs of Jezzine, carpeted in pine forests, vineyards, and orchards; the clumps of almond trees that cascaded down the hills of the Litani Valley. She paused at the panorama under the sentry of Beaufort Castle, sheer, inhospitable slopes plunging toward the river's churning waters — a sweeping view beautiful in its severity, like the face of a proud old man bearing the hardship he has endured. The road to Marjayoun was now paved. The town seemed smaller. For a moment, she could claim no part of it.

Then, as the car turned into Hayy al-Serail, she remembered her neighborhood. When the driver stopped, she and her sister tentatively walked up the hill to what was once their home. It felt familiar but distant, like an old picture in a frame, glanced at but rarely gazed upon. The joy of her return was shadowed by memories, the pain of her departure. Two women stood at the summit. One was Bahija Abla, her head wrapped in a scarf and her back arched like a crescent; the other was Bahija's sister.

"Hamdilla ala salaameh, ya banateh!" *the two elderly women shouted, sobbing.*

Raeefa rushed ahead, Nabiha behind her. In a moment of hesitation, she stood before them. Then she embraced the woman she thought was her mother. But no, she was her aunt. After an embarrassed pause, they all erupted in laughter and tears.

Raeefa shouted, "Am I crazy?"

As a young woman in America, Raeefa had declared a vow: If she ever saw her mother again, she would visit the Convent of Our Lady of Saydnaya, perched in the mountains beyond Damascus and renowned for its miracles. On that trip in 1960, after leaving Marjayoun, she fulfilled it. With a smile, she handed a donation to al-Hajja Maria, whose name Raeefa had scribbled in Arabic in her notebook before she left Oklahoma.

I went to Shibil's house on a Tuesday, the day we had planned to visit Suq al-Khan, the nearby market that convened on this day every week. Once inside, Shibil was typically askew. I had recently had lunch with Dr. Khairalla, and Shibil asked me where we had eaten. The Road Runner, I told him.

"What did you have?" he asked. "Coyote?"

He brought up the subject of women in Marjayoun.

"They don't have pretty eyes in Jedeida," he told me. "Like the eyes of wild cats."

What was really on his mind, though, was the hopeless congestion in his head. He showed me, tapping his forehead with his index finger. When he eats, he said, it echoes.

"Crunch, crunch," he said, moving his mouth in exaggerated bites.

Like any market, Suq al-Khan took its name from the lodging it once provided for caravans plying the roads that tied together Hasbaya, Rashaya, Kawkaba, and Marjayoun. The souk was most hectic in the morning. Shiite butchers shared space with Christian salesmen from surrounding towns. There was a charity box for Hezbollah, painted blue and yellow. On the windshield of a yellow van, a poster hailed the military: Salute to the Heroic Lebanese Army. On one table were cheap gold-colored ashtrays emblazoned with the imagery of every sect: Mohammed, Allah, bismallah al-rahman al-raheem, and pictures of Mary, Joseph, and Jesus, along with the Last Supper. Reflecting the region's demographics, the majority of the merchants were Druze, the only sect in Lebanon distinctive in their dress, at least in more rural, traditional places like Suq al-Khan. Their baggy pants hung loosely at the thigh, narrowed at the knee and shin, then came to a tight band around the ankle. They usually wore a white knit cap or white head-scarf. Druze women also wore scarves that covered the lower half of their faces.

The market was popular, bereft of any of Beirut's glitz, and distinctions between people blurred for the sake of commerce. The market's pride came from its wide selection of goods. Druze gardeners sold trees — plums, apricots, olives, pines, oranges, and lemons — planted in empty suet tins. Clothes dangled from the rickety frames of stalls. Others were displayed on tables, usually wrapped in the plastic that bore the mark of their origin — China or Syria.

Down the stair-like arcade, socks were sprawled across a red plastic table. Underwear was displayed on a blue plastic table with one leg made of wood, like a prosthesis. Stockings hung from green plastic twine; the most distinctive pair had candy-cane stripes. There was the utilitarian — gardening tools. And there was the tacky — one table overflowing with cheap jewelry, perfume, and cosmetics. At the top of the hill were two falafel stands, the scalding-hot cooking oil swelling in a black vat. The ingredients sat in bowls spread along the counter: tomatoes, turnips, lettuce, cabbage, onions, parsley, and French fries.

"Beautiful prices! Beautiful prices!" vendors cried out. Amid a jumble of black military boots, another merchant shouted at passersby. "Come over! I'll be honored!"

Shibil bought a half kilo of pumpkin seeds for 3,000 lira, about $2. I didn't buy anything. I simply stood there in appreciation, as if witnessing a world in its passing. Nothing threatened Suq al-Khan, a place that seemed as prosperous as any I saw in southern Lebanon. Baptisms, like Miana's, would last as long as the faith. But on that morning I wondered whether the communities that gave rise to these rituals would endure. In some ways, they were remnants of a culture that was being extinguished, of a town that was dying but had somehow managed to survive until now, even without its milieu.

It reminded me of what Shibil had told me when we visited the market in Marjayoun: "They bring people here to bury them."

Only the ritual was left.

From the start, Shibil had disapproved of my decision to renovate the house. All these months later, he seemed no more impressed now that I was living there.

He pointed out what I had done wrong and what I should have done differently. Unlike Karim, who had tried to impress me with his taste, Shibil suggested that I had been foolish. "Why didn't you get Grohe?" he asked when he looked at the bathroom fixtures. Why didn't you make the entrance to the Cave bigger? Why aren't the doors finished yet? He pointed to chipped paint in the salon. He worried about the echo that bounced off the tile.

He nodded, then said, with a little malice in his voice, "You paid for it."

It wasn't praise.

"What day is it?" he asked when I visited after he returned from Beirut, where he saw his brother. Shibil was pretty high, which he readily acknowledged. He had smoked two joints, and after I sat down, he began rolling another. "I'm not into time, man."

The last time I had asked him about his brother, he said, "Better, *alhamdilla*."

This time, he didn't go that far.

"Okay," he said, staring blankly at the television.

He blamed the cancer on his brother's wife, a heavy smoker who had quit only recently. When she smoked, he said, she never opened the window. When he visited them these days in the house where he grew up, he could barely force himself to greet her.

"No, no, I didn't say hello to her," he said. "Well, I said hi, but she didn't reply. She just stared at the TV. Why should I bother, then?"

"Really?"

"Even if she said hi, I'm not going to say any more. Fuck her," he

said. "My brother and I talked. I never did talk to her. My other brother
came. She said hello to him. She turned off the TV." He shook his head.
"That's her, man."

The older brother was the only one of his siblings who treated him
well. That was according to Hikmat. "He's an angel," Shibil said. "He's
more than an angel. He doesn't bullshit. He's down-to-earth. If he does
something good, he doesn't talk about it. He used to shell out money
to me since I was in eighth grade. He paid for everything for me. And
he doesn't say it. If he helps someone, he wouldn't say."

His brother's illness was weighing heavily on him, I could tell.

As we sat in his house, Shibil seemed even lonelier than before. For
so long, he had his mother and his older brother, but one was already
dead, and the other would die too soon.

The television flickered in the background. The images were all
bleak: tedious meetings of politicians bent on war, a shooting at a
school in Jerusalem. It was the Hezbollah station, and the program-
ming had a martial air: fighters saying goodbye to their families, recruits training, and, of course, dying. A song by a leftist Lebanese
Christian, Marcel Khalife, accompanied the video: "In our heart, the
branch of loyalty. We will remain steadfast here."

"If there's a war, do you think it will be war *tahina?*" Shibil asked.
"Do you know what *tahina* means?"

"Total war," I told him.

He shook his head.

The next night, I sat in Cecil's house, in a room with a low-slung roof
of wood beams. I looked out his window. The weather had grown
quiet, but not tranquil.

"I don't know, Cecil," I said, "I really worry about what's ahead."

He agreed, but anxiety seemed to him to be pointless, even a diver-
sion. "I won't worry about it because I've been there so many times. I
don't have to worry about it anymore." He had grown so tired of the
news that he avoided watching it on television, a routine that nearly
everyone else in Marjayoun followed with religious ardor. I told him
about the rumors I had heard — of protesters preparing to fire on po-

lice and soldiers, of an impasse lasting months, maybe longer, of the possibility of civil strife. Some leaders in Beirut were speculating that the civil war had already begun.

"I like life here in Lebanon, but I've never felt so hopeless," I said.

Cecil nodded. We agreed more often than not. I wondered if he was mellowing, hoping this was not the case. "At the moment," he said, "it is the worst situation."

23

OH LAILA

ONE DAY IN MAY, a visitor arrived. His face was vaguely familiar, but as he approached, on a morning as clear as its predecessor was not, I recognized him; it was Najib, the brother of Abu Salim, the stonemason. As I showed him the house, he brought up the frightening story of Albert Haddad, the collaborator who had lived upstairs during the Israeli occupation.

By now I had heard plenty of stories about Haddad. Najib seemed to have the most authoritative account of him, and as the workers went about what I hoped were their final tasks, he laid out the details of what was almost certainly the darkest chapter of the history of Isber's home.

Najib insisted that Haddad, fluent in Hebrew, had worked with Unit 504, an Israeli military intelligence outfit, and that he had carried a pistol with a silencer, meant for assassinations. In Najib's telling, Haddad had run a network of three hundred informers, stretching from Jezzine to Kfar Killa and Qlayaa. As part of his payment, the Israelis had given him a white Range Rover, a green Mercedes, and a dark blue Toyota van. Aside from his vicious dogs, the bane of the neighborhood, he had only acquaintances; anything more would have required trust. Married twice, both of his wives had died of cancer, and he himself had escaped assassination in 1995. "He was a very dangerous man," Najib told me. "Everybody was scared of him."

I asked Najib where Haddad, a native of a town near Sidon, lived these days; I had heard he had returned to his birthplace. Some had told me Israel. Another friend said that Haddad was in Kiryat Shemona, the Israeli border town in the Hula Valley and the scene of a massacre by a Palestinian splinter group in 1974.

"No one really knows," Najib said.

When he asked me whether Haddad had damaged the house, I thought of all the scars in the *liwan* and the garage, the injustices against the arches and wood-paneled roof. I nodded.

"Don't worry," Najib told me. "Many things worse happen in Lebanon."

By the time Bahija had turned ninety, her back was disfigured by osteoporosis, and her face and hands bore the years of Marjayoun's sun and wind. The photograph I gaze at reads El-Marj, *Marjayoun's venerable studio, and in the picture my great-grandmother wears black. At the time she sat for this portrait, Bahija was already changing. New acquaintances had become a blur, consigned to anonymity. She also forgot familiar names. Ali and Hussein had to remind her of requests she had made the previous day. The wanderings of her imagination — of Isber returning from the Houran, of a house still full, of a family still together — seemed to have become more real than her actual existence. Bahija Abla's house remained hers until only months before she died, in 1965. It was around this time in Oklahoma that Raeefa's relatives noticed that Bahija's daughter was looking piqued and wan.*

"Well, I've had this flu," Raeefa explained.

Her youngest son was a doctor, and Raeefa heeded his advice to get a checkup.

At the age of fifty-seven, she learned that she had stomach cancer.

"In those days," her oldest daughter recalled, "it was a death sentence."

Raeefa never accepted it. She had struggled through tragedy, a life that was served rather than enjoyed, to reach this point. Her children's prosperity continued. Grandchildren crowded her house. She would not die; it was simply out of the question. Not once demanding sympathy, she willed herself on.

I have a picture of Raeefa dated November 28, 1968. In the photograph, her fourteenth grandchild, just two months old, sits in her lap. Raeefa remains an elegant woman. Twice a week she visited the hairdresser. Yet ravages of the cancer and her surgeries are evident. Clearly, she was suffering. According to her children, she rarely slept well. During the day, she sometimes sat in the corner of an orange sofa, her legs pulled beneath her. She wrapped herself in an afghan then tucked it in the cushions. Rocking back and forth, ever more emaciated, sustained by her will alone, she waited for her youngest son to find someone to save her.

In those quiet moments, Raeefa recited the nursery rhymes she was taught as a child in Marjayoun.

Oh Laila, oh Laila.

She shook her head, her eyes lost in thought. Memories fumbled behind shadows, hinting at what they once were. Again, she was leaving home.

Oh Laila, there are no eyes like her eyes,
And the magic in her eyes,
Oh Laila, oh Laila . . .

"Why is it that I'm remembering these songs I used to sing when I was a little girl?" Raeefa asked her daughter.

The children of Bahija who stayed in Lebanon cared for the house in Marjayoun after she died. They checked on it. They visited, especially in the summer. For years it resembled the place that Bahija had so arduously cleaned, even if it was no longer a home.

Ten years after she died, in April 1975, the first shots were fired in a civil war that would rage for fifteen years in Lebanon. In the south, too far from Beirut and too close to Israel, the Lebanese army splintered. Palestinian factions multiplied, springing forth from their stronghold in the Arqoub. Residents in places like Qlayaa banded together with arms, and Syria and Israel courted their clients. Marjayoun, inevitably, collapsed.

A Christian cleric appraised the situation this way:

Paralyzed municipality. Electricity and waterlines badly damaged. No telephone. District hospital closed (there are only four nuns/nurses in the hospital). There is only one doctor (old Dr. Shadid, who does not have equipment or medicine). Only one dentist (Dr. Karbis). Only two schools operating on limited, part-time scale. The Marjayoun National College closed. The Serail building closed . . . No courts. No police (the few gendarmes joined the local army). No postal service. Most grocery stores closed. There is a desperate need for commodities as well as for cigarettes and beverages. Need for fair quantity of cement and other building material in order to repair war damages. Bank is closed. There is no commercial activity. Agriculture totally paralyzed. [Quoted in Beate Hamizrachi, The Emergence of the South Lebanon Security Belt]

Bahija's house fell victim, too. Looters from the Christian town of Aishiyya plundered the house, taking everything they could carry. The rest they dumped in the garden, near the patch where Bahija once grew tomatoes. Ten pieces of furniture — couches and chairs crafted of walnut wood, mother-of-pearl, and camel bone — were strewn in the dirt. They stayed there, ruined by the sun, wind, and rain, in a garden overgrown with weeds, wild lilies, and the fuchsia blossoms of the four o'clock flower, a motif of old and abandoned houses in Lebanon. Mortars followed, gouging a crater in the roof above the salon. For two years, rain fell on the red carpet, laced with cream and shades of blue, that Nabeeh bought for fifty gold pounds in Damascus. The marble floor that Bahija cleaned each day of her life no longer shined. The purple, green, and yellow patterns of the cemento tiles were lost beneath dust. Pieces began falling from the wooden tracery of the triple arcade. The front door, its finely wrought wood mimicking a ruffled sail, was shut with a rusted padlock. No one lived there anymore.

• • •

It was Shibil's birthday. He had turned fifty-nine. When I arrived, he met me at the door with an unopened bottle of Glenfiddich wrapped in a black plastic bag.

"Oh, the good stuff," I said.

It was a step up from the Grant's that Shibil usually served, always generously.

"Of course," he answered. "Why not?"

"It's a celebration," I said.

Without smiling, he said, "The government should declare it a holiday."

My eyes locked on his clothes, an outfit I suspected was last glimpsed on an aging 1970s porn star at a California pool party. His short-sleeve shirt had three alternating stripes: dark green, white, and a light green that stretched across his ever-growing girth. His shorts matched the belly-hugging band of green, and his sneakers were another shade of the same color. He was, shall we say, in a festive mood, and as we climbed a steep road in his vintage white Mercedes, both of us spent most of the time laughing.

"I'm so close to finishing the house," I told him.

"Upstairs?" he asked.

"Yeah."

"Helter-skelter," he said.

The end of my stay in Marjayoun was approaching, and I had invited Shibil to have lunch with me in Shebaa, a town near a disputed, eight-square-mile patch of land at the intersection of the Israeli, Lebanese, and Syrian borders. These days, most people knew the area for its politics: Hezbollah had vowed not to surrender its weapons until Israel withdrew from nearby farms it occupied in the 1967 war. I knew it for its reputation as one of the most beautiful locales in the south, home to some of Lebanon's best cherries and a stunning spring called Naba al-Jawz. Fond of the restaurant there since his childhood, Shibil quickly agreed to come along, and, a little wistfully, I realized it might be the last time I would see him for a while. My leave from work at the *Post* was ending, and though I planned to return often, I wouldn't be living in Marjayoun for a while.

"Shibil, can I ask you a question, a personal question?" I asked as he drove. "Are you happy?"

"What?"

"I'll tell you why I'm asking," I said. "We talked about Assaad being depressed, and you don't seem depressed to me."

"I got a lot of crap going on, you know, in and out, shit like that. I want to get the best I can out of it. I'm not really happy, no." He sneaked a drink of the Glenfiddich. "Just passing the time. Like that. Being bored also. I don't seem to be very unhappy, but I'm not happy, either."

Silence followed, and eventually I said something to break it. "I think I like Marjayoun now. Maybe not the town itself. But my friends, the house, the setting."

"You don't think I have the same feeling?" he said.

"When do you think you were happiest?" I asked.

"In college."

"When do you think you were happiest in Jedeida?" There was a long pause, maybe a minute or so, as I smoked a cigarette. He never answered.

Shibil treated his car almost like a child. His only possession in the world, it figured into most conversations. As we continued on and the gas gauge neared empty, we pulled into a station to fill up. After that, we stopped at three more stations to find the transmission fluid he wanted. (Of course, we didn't find it.) Along the road, he inquired about a tire. Then, while chugging up a hill, we drove past a black goat that alarmed him. As always, he was superstitious.

I mentioned Dr. Khairalla to him. He had seemed really sick when I saw him earlier in the day. "The cancer has spread to his back," I said.

"No shit? From the prostate? Can he cure it?"

"I don't think so."

"I pray for him," Shibil said. "*Wallah al-azeem*, every time I go up to Mar Elias I pray for him. I pray for my brother, I pray for Dr. Khairalla. Because I love this man, *wallah*. He's a good man, a good man. If there's another like him in Jedeida, great. We'd become a good town. But there isn't. He's one of those people who are very scarce."

Shibil pulled his Mercedes into a street crowded with families, their cars parked at the curb on both sides. Even before he turned off the engine at the restaurant, I could see it in his face. He was disappointed. The music was awful Arabic pop, performed live at deafening volume. We sat under a walnut tree, a *jawz*, which gave its name to the spring.

"A bunch of fucking kids, man," Shibil said. "I thought it was completely different. When we used to sit here, you could smoke a joint. I don't know what happened." He paused. "I wish I could see some of my old friends."

The waiters brought us three plastic cups and a metal container of ice with tongs. Shibil had a black plastic bag full of raw almonds, whose shells he threw over the wall of a garden. He filled my glass with ice. "*Shou, ya ammee,*" he barked. "*Hutt whiskey!*"

"I'm going to miss you, Shibil," I said.

Shibil often grew excited when someone showed him kindness. "I'm going to miss you, too!" he shouted. "For Christ's sake! When are you going to be back?"

We toasted as we drank. "Cheers!" We clinked glasses. Then, cheers to my daughter Laila, and our glasses banged again. And cheers to Abu Laila, the father of Laila, and more toasts followed. I smiled. Shibil said he was still determined to buy me a blue eye "to repel jinxes and shit," which he had first mentioned back in January 2006.

The grilled chicken and lamb arrived, and Shibil started reminiscing about his days in Oklahoma working at Sears. He told me the first time he got high was in 1973, at a party with his friends from work at the Skirvin Plaza. "I loved it," he said. At work he was popular, if lazy; "King Skate" was his nickname. His college transcript read "Sam." So did his social security card. I started thinking that Sam was, in fact, another person, even to Shibil himself.

"I have all the phone numbers of the girls I used to date. Well, most of them."

"Where do you keep them?" I asked.

"On a card with other papers. I once counted how many I dated."

"How many?"

"About one hundred fifty-four."

"No, you didn't. Really?"

"Yessiree, maybe even more."

I asked Shibil why he hadn't stayed in Oklahoma.

It was 1975, he said, and on his nightstand he had a portrait of his parents and a picture of Karen Chase, from Jones, Oklahoma — "the only girl I loved in the States." One morning, he woke up and her picture had fallen down. "I decided to go back to Lebanon."

"Why?"

"I'm superstitious," Shibil said. "If my parents' picture had fallen, I probably would have stayed in Oklahoma. I don't know. They are just signs." He returned on March 18, and the civil war started on April 13. "Tell me how lucky you can be," he said.

"Do you regret it?"

"No," he said. Then, in that peculiarly Shibil way, he changed direction. "I do, but I don't want to put it in my mind."

Back in the car, both of us a bit drunk, Shibil shifted into first, grinding the gears, and we chugged forward.

"I wish I had a small vacuum cleaner to suck on my nose and drain all the mucus out," he said, pointing to his head, his congestion still dogging him.

Then, tipsy, he told me a joke as we barreled along. "What's better, Three Musketeers or Milky Way?" I shook my head. "Chocolate," he said, laughing heartily.

Only Shibil understood.

24

MY JEDEIDA

I T FELT AS IF summer had arrived, bringing not just the dead-line for my finishing the house but the end of my time away from the newspaper. It was May, and I was slated to return to work by the end of June. I was trying to forget these imminent events, spreading dirt in the garden that we had brought from the valley. As always, the work was meditative, though Abu Jean was not.

Sitting in a white plastic chair that he had conveniently pulled near me, he slowly smoked his Cedar cigarette and gave me orders.

"The dirt's low over there," he shouted, pointing. "Put more!"

And so it went, so reflexive I suspect he knew not what he was doing: There's too much dirt on that mound. Turn the wheelbarrow around while you fill it. Why don't I wait until the dirt is drier so I can break the clods easier? Add some dirt over this pipe.

"*Ya ammee*, that's good enough," he said. "There's no need to work too hard."

"I work, Abu Jean, and you sit," I said, laughing. "How is that?"

He shook his head, his reflex when he didn't hear something.

Across the valley, over the town of Khiam, clouds gathered on the horizon, the gray muting a sunlight that had felt especially soft for the spring. Hardly a car passed that afternoon, making the neighborhood quieter than usual. Down the lonely street, Joseph Abu Kheir, the painter, approached. "There might be rain," he told me, smiling. "It's going to be good for your plants." It was the day when I began to feel the end of things. I knew I was bracing to return to a different world.

Reality intruded, even in this small town. Text messages burst onto my cell phone as clashes reminiscent of the civil war erupted in Beirut. Over lunch at Toama's house, the television delivered a barrage of news. Toama's son, Alaa, was studying in Beirut, and twice Toama called him to see if he was in danger. Everyone in Marjayoun seemed to be watching TV; it was as if the clashes had broken out next door. The fighting that punctuated Beirut began every conversation. Voices expressed worry, and the scenes of burned cars in streets raked by gunfire replayed again and again in my head. War was coming. Again the lull was over.

For an hour or so I fought the guilt — or perhaps a renewal of the old ambition. I should be in Beirut, I thought, working as a journalist, but another part of me was so wary of that old life of guns and misery. I did not want to see Tyre again, or Qana, or Baghdad. I wanted to do nothing more than move dirt from one place to another.

And then I left.

"*Intabih, intabih. Beirut kharbaneh,*" Abu Jean said over and over. Be careful, Beirut's a complete mess.

I took the same road I had taken scores of times, though on the day of my return to reporting the late dusk caught an especially beautiful light. Along the coast, it changed. Knots of soldiers parked their generation-old green tanks on the curbs. Their helmets were vintage, though an occasional gunner behind a turret wore the more stylish green beret. White and red barricades were tossed haphazardly in the streets, and the acrid smell of burning tires had drifted from roadblocks still far away. I passed a bridge destroyed in the 2006 war, the crisscrossed steel that reinforced it spilling out. Pieces of cement dangled. From half an hour away I could see Beirut, filled with fears and whispers of impending civil war, but always with the look of a picture postcard.

The opposition, led by Hezbollah, had called a strike, ostensibly to protest the deteriorating economy. But it was really a raw show of force directed at a government that Hezbollah believed threatened its armed wing, what it called the Islamic Resistance.

As I drove into Beirut, things had already taken a turn for the worse, with hundreds of supporters of Hezbollah and its allies blocking the roads. Clashes erupted in stages; neighborhoods flared like torches, their flames igniting streets with both Sunni and Shiite residents. "God is with the Sunnis," backers of the government cried. "The Shiite blood is boiling," their opponents shouted from across the road.

Escalation prompted reprisal, and provocation inspired vendetta. Government troops in armored personnel carriers raced through neighborhoods trying to contain the fighting and disperse crowds. They could only shoot in the air. The city was soon paralyzed; trucks and bulldozers dumped heaps of sand on the road to the airport, where nearby rioters streaked through on scooters.

"Those who try to arrest us, we will arrest them. Those who shoot at us, we will shoot at them. The hand raised against us, we will cut it off," vowed Hezbollah's leader, Hassan Nasrallah. The government's decisions, he said, were a declaration of war. "Our response . . . is that whoever declares or starts a war, be it a brother or father, then it is our right to defend ourselves and our existence."

Hezbollah's men and allied fighters soon deployed across mainly Muslim west Beirut, routing in just hours militiamen loyal to government figures. Masked men armed with assault rifles roamed shuttered streets amid smashed cars and smoldering buildings. In ammo vests and black baseball hats, they stopped traffic at checkpoints, demanding identity cards. The television and radio stations of government supporters were forced off the air. More barricades were erected on highways and at intersections, closing the airport and the port, and dividing neighbor from neighbor. Even the opposition's supporters cringed at the sight of militiamen sipping coffee at Starbucks, their rocket-propelled grenades resting in chairs in a distinctly Lebanese vision of globalization.

All the time, I was watching with the clock ticking. A few weeks before, I had promised Laila that I would take her to my mother's wedding in Washington. She was the flower girl, and she had already picked out a white basket with pink petals, black patent-leather shoes, and a pink skirt with matching ribbon and flowers stitched in the middle. I had disappointed Laila before. How many times had she been betrayed by my career?

Nearly five years before, on December 14, 2003, I had arrived home in Washington from Baghdad, picking Laila up from the house of her mother. The next morning, my daughter and I had sat together on the couch, waiting for the start of a Green Bay Packers game. Then my phone rang. American soldiers had captured a bearded and haggard Saddam Hussein, who had been hiding in a six-foot-deep pit at a farmhouse near Tikrit, not far from his hometown. My editor never asked me to return to Baghdad to cover the story. He didn't have to. "I'll leave it up to you," he said simply, and I knew what I was supposed to do. I returned Laila to her mother's house and left for Baghdad. I had the awful feeling that I was destined to disappoint her again and again.

This time I was not going to let my daughter down. My flight to Paris was at 4:15 A.M., and after waiting for a few hours at my apartment in Beirut, a colleague came to drive me to Hariri Airport soon after midnight. I had stayed long enough to write a story, and I worried about whether I had given myself enough time to catch the flight. The wedding was on the weekend, and I had a day or so to get to Washington for the ceremony. Beirut was somber and somnolent. Usually rollicking at midnight, my old neighborhood was deserted, all lights dimmed. A few cars passed on the streets, but for the most part it felt like the hour before dawn. The road to the airport was strewn with the roadblocks from the day before. Refuse from overturned trash bins smeared the asphalt. Smoke rose from wood and trash that was still smoldering. From the car I saw a knot of young militiamen on the left side of the street. They chatted idly until they saw our car. *Something to do!*

One of the men swaggered over to the car with an automatic rifle. Another approached the window. Where are you going? they asked my Lebanese colleague.

"I'm trying to get to the airport," she said.

"He is traveling?" the man at the window asked. "He can't travel. You can't get there."

"We're going to try anyway," she said.

He grew angry. "Why? The airport's closed and the airport road's blocked," he said. "See, you are lying to me."

"No, I am not lying," she said, not losing her composure.

"If you tell me you are going home, I will believe you and show you the way. But you are telling me you are going to the airport, which is closed, and the road is blocked."

"We're just going to try," she said.

Growing bored, the militiaman let the car pass. The same scene was repeated at two more checkpoints. At a final roadblock, three pieces of iron scaffolding strewn about, I got out of the car and moved them.

The airport was a grim scene — empty but brightly lit. There was one other passenger in the terminal, and I could only guess how he had arrived. I looked up at the television screen. I was determined to get back to Washington. But no: *Canceled* was written next to every flight. I stared at the screen for a few minutes, hoping for change, then lay down on a steel bench and propped my head on my computer bag. Every half hour or so, speakers carried a recording played, in succession, in Arabic, English, and French: "Welcome to Beirut's Rafik Hariri International Airport. We are pleased and honored to have you in our city."

At 7:30 A.M., I woke up to thick black smoke billowing over the airport. Within an hour, protesters had ignited two more piles of tires at its entrances, and the acrid smoke left a fine black gauze on every surface. Everywhere there was a panicked sense of siege. By midmorning, my spirits low, I gave up. I decided to cut my losses and return to my apartment.

Text messages kept arriving. *Exchange of gunfire taking place in al-*

Madina al-Reyadeya area in Beirut . . . Beirut port closed . . . Saudi Arabia urges a conference of Arab foreign ministers . . . Heavy gunfire near Beirut Arab University and Cola in Beirut.

I didn't want to pay attention anymore. Numbly, I walked from the airport entrance down the street as the bright sun baked the ocher-colored landscape. Pulling my duffel bag behind me, my computer bag over my right shoulder, I arrived at a roadblock, interrupting what looked to be a party for a group of young men. Some wore masks; others had tied scarves around their faces. A few darted around on mopeds, omnipresent in wartime.

Flames poured out of a nearby trash canister, pushing smoke skyward where it mixed with more smoke. Idling taxis charged 10,000 lira (nearly $7) to go from the roadblock to the airport, a drive of a few minutes. Boys demanded 1,000 or 2,000 lira to help carry anyone's bags down the road. As I finally moved through the barricade, someone called to tell me a Middle East Airlines plane would be taking off. Someone else phoned to say I could get a connecting flight to the Persian Gulf and Europe, including one to Paris at 4 P.M.

I turned around and headed back.

My shirt was soaked with sweat, from nerves and exhaustion. My shoulder, the one shot six years before, hurt. As I walked to the airline terminal, toughs loitering in the street kept approaching, most of them harmless. Clambering over a dirt pile, still dragging my duffel bag behind me, I stumbled, then caught my footing. Rivulets of water trickled down the street, the asphalt cracked and worn. Smoke made my eyes water, just as I glimpsed the tranquil Mediterranean in the distance.

One plane managed to leave the airport that afternoon.

I was on it.

By the time I returned to Lebanon, scores had died in the paroxysm of violence, the worst fighting since the end of the civil war. Each side had its martyrs, whose memories would be manipulated by the so-called leaders who had stage-managed the bloodshed. It was probably only the decisiveness of Hezbollah's victory that stopped more from dying. Somewhat cynically, the country's lords soon came together to deliberate, as the government rescinded decisions that Hezbollah had found so threatening. Arab mediators then invited the factions to meet in the Persian Gulf emirate of Qatar, an announcement delayed several times as the politicians haggled over words of the communiqué that announced the dialogue, which took place almost completely on Hezbollah's terms. After five days of negotiations, which repeatedly verged on collapse, a deal was struck to end the crisis and choose a president. To me this finale seemed more respite than resolution. Nothing ever seemed to be resolved here. Lebanon's dramas, I thought, were simply too big for its small stage.

On a sunny morning, the curious and the committed came to see the end of an eighteen-month sit-in in downtown Beirut that the Hezbollah-led opposition had organized, paralyzing part of the capital.

Workers directed by Hezbollah cadres carrying walkie-talkies — no doubt the same marauding gunmen who had fought only weeks before — began removing worn mattresses, soiled pillows, small butane stoves, leather couches, pots, pans, rusted tent poles, and cheap Syrian-made heaters. Wearing yellow caps, the movement's men planted roses, bushes, and trees in the once manicured gardens. Call it a cosmetic touch on a scarred city.

As I walked around Beirut that day, I passed an electronic billboard that read *1,193*, the number of days since the former prime minister Rafik Hariri was assassinated in 2005. It was his death that had initiated the crisis — at least this latest one — and I felt the same anger that I had heard so many people convey to me. It had been three years, three months, and eight days of crisis, tension, apprehension, anxiety, and unease. People had died. Lives were shattered and ruined, for nothing.

Three years, three months, and eight days.

The day before my return to Marjayoun, I called Cecil's friend the eighty-three-year-old architect Assem Salam and drove over to see him at his grand villa in the Beirut neighborhood of Zqaq al-Blatt.

Assem was simply tired of crisis. Like me, he felt frustration at a country blessed with talent and resources that seemed forever inclined to civil war, occupation, and force of arms, the reflexive instruments of change since the darkest days of World War I. I mentioned how I felt, that Lebanon was too big to be small, overwhelmed by conflicts that perhaps deserved a domain bigger than the country itself. He shook his head. It was too small to be big, he insisted, and the traumas were of its own making, forever stunting its ambitions of becoming something greater.

"Is Lebanon really viable?" Assem asked.

I shrugged. "There has been a question mark since the inception of Lebanon," Assem told me, "and that question mark remains."

We sat for a moment in silence, and he puffed on his cigar. Light reflected faintly from stained-glass windows of red and blue, under graceful Levantine arches built 176 years before, when frontiers stretched much farther. Everywhere were vestiges of a more confident age.

"I wish I had been born in Syria," Assem said. "Or in Egypt. Can you imagine living in a country that has gone through thirty years of this? What kind of country is this?"

He shook his head, his anger giving way to dejection.

"There's something wrong here," he said, "something wrong. You have to ask yourself."

The next morning, I left my apartment in Beirut and returned to the house in Marjayoun.

Ya maalimeh! Fadi shouted at me.

I smiled. Nearly a year after starting, I, too, was addressed as a *maalim*.

"The floor's going to be the mirror you use to shave," Fadi told me as he rolled his hulking cleaner over the marble of Bahija's *liwan*.

I nodded in approval. For months, the marble had been buried in wood, dirt, sand, and stacks of tile, the rusted barrel, with swampy water, its sentry. Who would have recalled all those days of Bahija's scrubbing on hands and knees. Now, after four decades, the floor shined as it once had. One more scrub and Fadi promised that I could go ahead and put the razor to my beard.

"Enough talk," Abu Jean barked at him. "You should work as much as you talk."

"You could use a scrub, too, Abu Jean," he said.

With only a week before my leave ended, the drive to finish the house hurtled forward as quickly as the culmination of the crisis I had left behind. For perhaps the first day since the project started, everyone was at the house, working away in a scene that was remarkable for all its interlocking parts. Fadi continued in the *liwan*. Ramzi was finishing the roof, finally. Toama reworked the gypsum of the salon's old ventilation system, those hand-wrought designs that in summer brought cool air into the house's largest rooms. Paint was applied to the steel beam holding aloft the balconies, still bearing the inscription of the French company that delivered it nearly a century before: *Senelle PN 180*. Even the itinerant blacksmith stopped by to take measurements. Parquet went down on the floor, shutters went up on the windows. Charming but fickle, Cesar, in charge of the windows, promised absolutely, without question, that his work would be finished Sunday. "Monday at the very latest," he added. Malik and I professed our affection for each other. Even if I was still intimidated, I could offer praise. "Your father is the best *maalim* in Jedeida," I told his son, Nicola, who joined him at work. Malik beamed, declaring he was here only in fealty to me. "I'm working for Anthony from the heart," he said.

These days, even crises were resolved. Malik had sent measurements of the black granite for the kitchen counters to a workshop in Khiam. That evening, the head of the workshop came by, somewhat furtively, and took his own measurements. They were predictably wrong. When the granite arrived, I knew this solely by the shouts that echoed loudly from the kitchen.

"Who dared change my measurements?" Malik cried to no one in particular. "Who would dare do that? I've been a *maalim* for twenty-eight years and someone is going to question me?" He said it over and over: "I've been a *maalim* for twenty-eight years."

He turned to me, still shouting, as if expecting me to both confess and console. "I can't install this. It's impossible. I can't do it," he declared before throwing up his hands.

It meant that we would have to recut the half dozen or so slabs. The granite in hand, Malik piled in the car with me and Abu Jean in an ordeal that I was sure would take days, which of course I didn't have.

I would beg, appealing to their sympathy. They would demur, citing their backlog. Our order would fall to the end of the queue, and the kitchen I had hoped to finish this evening might get done next week but probably would take much longer. In a sense of the preordained, the stonecutter was sleeping when we arrived at the workshop in Khiam.

"Get your ass out of bed!" Malik shouted at him.

He did, his eyes still half shut, and Malik gave no opportunity for hesitation. In a stupor, he stumbled to the saw, and Malik barked out orders, measurements, and recriminations. "Twenty-eight years!" Malik yelled.

The saw drowned out any other noise, as water, keeping the granite cool, flew into the air. "That's right, that's right," *maalim* Malik said, pleased as the stone was cut to his specifications.

One by one, Abu Jean and I carried the newly cut slabs to the car. Fifteen minutes later, we were driving back to Marjayoun. An hour more and the kitchen would finally be finished.

Over these days, I stepped back to appreciate what we had accomplished in a year. The detritus of the house's construction no longer bothered me. I looked past the cardboard on the floor, the buckets filled with tools next to the windows, the empty packet of tile emblazoned with *Ceramic Cleopatra*, the empty liter bottle of Pepsi, the five plastic cups, a Fig Newtons wrapper, and a scrap of newspaper from 2003 pinned under the cinderblock. I ignored the three ladders, the garden hose that snaked across the marble floor, the three buckets of paint, the six cardboard boxes, the tube of caulk, the hammer, the masking tape, and the piece of dismantled scaffolding. The entrance was yet open, letting ants and mosquitoes amble through the house as they had for years; the same breeze still ventured through the rooms unhindered. Yet I saw what had become a home, and its audacious beauty had emerged. We were almost done.

A few days later, it was, or as close to being completed as it could be before I left.

I walked up the stairs, their chips and cracks recounting their age, the stones speaking of time and endurance. Everywhere the utilitarian had become elegant. I entered the nearly century-old door, its façade like a billowing sail, and walked inside the *liwan*, bathed in light pouring through three arches. Beneath my feet was the marble Bahija had cleaned, bordered in black. Four small squares in the center were the same color, a suggestion of delicacy. The tile that Isber Samara had purchased after World War I mixed with the tile over which I had tepidly bargained for with Abu Ali. In form, the ceiling was the same as it was when Bahija lived here alone.

On the balconies, which looked out at the world, the iron railing remained part of the house. Built in the traditional Eastern way, the *darabzin* had not a nail in it. And no welding. Each piece was fastened with another piece of iron, small and wound like a clasp. In a way, they were built for age and perseverance. A welded piece would snap someday, pulling the balustrade itself apart. With the clasp, the grill could age more gracefully, pulling and bending as it gathered years, without breaking. I saved old cemento tiles and returned them to the house's entrance. Their array of shapes and colors made them three-dimensional.

As I walked through it all, I had a sense of belonging, and I felt an affinity. A year ago, when I came with my cousins, we whispered, fearful of raising our voices in a house estranged from us. Now I was alone, and it was quiet. The silence felt to me like acceptance.

One afternoon, as I prepared for my departure, the priest came by. I had never met him before, and after he left, I forgot his name. Burly, with a black beard, and dressed in black, he was known in town for making mosaics. He had a few suggestions, he said.

Could I make the stairs of wood instead of iron? Could I change the cords from which the lamps were hung? Why didn't I use wood for the shutters?

"The work is nice, but *harram*, what a shame, it could have been better," he said.

He shrugged his shoulders, humorless, as he sauntered off, offering no blessing.

I walked outside afterward and my neighbor approached, the house heaving toward its completion. Why, he asked, didn't I make the shutters green, the same as those on the house next door, where his brother-in-law lived? I liked brown better, I told him.

Like the priest, he shrugged. "You spoiled our house," he said. "Anyhow."

Their words didn't matter, though. The house was mine.

Back in 2003 in Baghdad, I told Abu Jean, Saddam Hussein had called the war that the Americans began the *maarakit al-hawasim*. It meant, roughly, the Decisive Battle, a phrase that soon became ironic.

"Today is our *maarakit al-hawasim*," I told Abu Jean.

He nodded, not comprehending.

We had a window of a few hours, maybe less, to finish what we could finish before the movers brought in my furniture. Here is what we had to do: Fadi wanted to put a final polish on the marble; the blacksmith was trying to solder the iron stairs; Camille was supposed to adjust the wood tracery in the arch over the door; Toama had to touch up the paint; Emad was screwing in the light-switch covers; and the rest of us were sweeping the floors, mopping the tile, hauling away the trash, spray-cleaning the stone with water, and, barefoot, with our pants rolled up, shoveling whatever drained off into buckets.

The movers soon arrived, and as I gazed inside the *liwan*, I realized that for the first time since I had come to Marjayoun, something unusual had happened: Our work was done on time. The floors were barely dry as they hauled in the furniture from my Beirut apartment, where my lease had expired, leaving me without a home there. The marble glistened like a mirror; in its reflection I saw my beard, growing ever grayer.

As the evening ended, I sat on Isber Samara's balcony. I had prepared a meal with as much as I could harvest from my garden — onions in a dish of lentils and rice, green tomatoes, *miqta* that I salted just right, and a salad with a pepper, mint, onions, and yet more *miqta* that I had picked. I took my place at a table whose marble top was taken from part of the kitchen counter on which Abu Elie, the squatter downstairs, had once cut parsley and tomatoes. Beneath my feet was a pattern of more tile that I had bought from Abu Ali. I looked around the house, at the arch, the door, the shutters, and the stone, and I felt something that was always fleeting: satisfaction. With a little more money, I could have bought prettier handles for the windows. With a little more time,

I could have saved some of the old doors and arches. But all in all, I had turned an abandoned house, disabled by war, into a place that exuded a kind of peace. Rather than just a channel to the past, or a facsimile of it, it had become new, part of what was and what would and could be. Isber's home, born of ambition, had been burnished by the sacrifice of two parents who chose safety for their children at the cost of their own loss. It was a place where my family could take what they needed from the past, as I had, seeing in its stories the comfort I sought and the promise I found. Sometimes it is better to imagine the past than to remember it.

The next morning, a sunny one, I stood in the garden looking at the old doors that we had pulled out and discarded months before. I stared at some pieces of marble from an old countertop. I was tempted to throw them out. Then I remembered Raeefa in Oklahoma, hoarding half bricks, saving pennies, holding on, making everything count. I knew I couldn't throw these things away. I would save all of my family's lives that I could, every fragment from Isber and his time, every piece of the past, everything my great-grandmother and grandmother had touched. In the house in Marjayoun I could see the past in the present, see the things worth preserving. The Levant is no more, but I had been reminded — by the grace of the triple arches, the dignity and pride of the *maalimeen,* the music of Dr. Khairalla, and Isber's sorrow and sacrifice — that behind the politics there were prayers still being said with hope for what draws us together.

There was a Jedeida today. Then there was a Jedeida that we remembered, or imagined, or wanted to imagine, filled with friends and relatives, houses that embodied a forgotten past, glimpses of Mount Hermon, and a reflection of ourselves or what we wanted ourselves to be. My Lebanon was my grandmother's, a place besieged by war, but my Jedeida was an idea born of Isber's house. Nothing could wreck it; no war could destroy it. I could always go there. It was always with me.

I promised myself that I would save a jar of olives for my last night, to celebrate with when the house was finally done. Taking it from the cabinet, I put a dozen or so in a white bowl fitting for my ceremony. I tried two or three. They had aged well despite being picked too early by a novice, a newcomer, but they were no longer bitter and the taste of salt had been somehow subdued. As I ate them, I thought of the day when my daughter and I would savor the fruit from her tree, the tree planted the day I began the journey back to Isber's. There was more life to come in this old house.

It was the start of a fine day in June. I was standing on Isber's balcony when my phone rang. It was Dr. Khairalla.

"I'm not doing too well," he told me after we exchanged greetings. "I'm staying in bed."

There was a hint of irritation in his voice, anger directed at himself, or perhaps at his body for failing him when he had so much to do. The cancer, he told me, had spread further along his spine, from the L4 to the L5 vertebra.

I felt uneasy, so I changed the subject.

"You wouldn't believe the passiflora, Doctor," I said, looking down at the garden. "It's just thriving."

He had given me the passionflower, and soon after I planted it, it sprang forth, hungry for life. Its tendrils had soon wound their way up an iron fence. Sometimes it seemed to grow across a stone in the wall in one day.

"You should cut off the top so that it climbs in both directions," he told me.

I could hear a glimmer of interest in his voice, even excitement. "Has it flowered yet?" he asked. "No," he added almost immediately. "It's still too early."

The next day, I went to see Dr. Khairalla at his house. Ivanka answered the door. He was still in bed, she told me, and couldn't come down. She suggested I go up; his was the last bedroom on the left. When I saw him, I tried to take in how much he had deteriorated in the two weeks since I had last visited. He was lying in a white wooden bed, swallowed by the blue plaid sheets and white comforter. His face was wan, even sallow. As I entered, he struggled to pull himself up to a pillow that was propped against the headboard and covered in a white towel. On the walls around him were five pictures of his two children and his grandson Jean. Against the mirror on the nightstand was an icon of the Virgin Mary with child. Next to it was a single red rose in a white vase that he had picked for his wife on her birthday, June 19, the same as Laila's.

As we sat there in silence, the television played in another room. Ivanka was watching the beatification in Beirut of Yaàcoub Haddad, a Capuchin priest from Lebanon, and I heard snippets of the ceremony: "Pray to God . . . as a symbol of our faith and love . . . thank Him for what He gave."

"Have you seen the passiflora's flower yet?" he asked me again.

I shook my head. He gingerly got out of bed and took a book off a shelf, *The Complete Indoor Gardener.* He sat down and opened the book to the index, looked up the flower, then turned to the page for me. There was an exquisite picture, and for once I could grasp how Spanish missionaries saw in it the death of Christ — its sepals and petals as disciples, its rows of blue and purple as a halo (or crown of thorns), and its stamens and styles as the wounds and nails of the cross. So subtle and bold, the flower felt animated.

"Next year, yours will look like this," he told me.

For the first time, he didn't offer to help prune the trees. He didn't suggest what we might find in the Jibchit nurseries, and he didn't propose to bring me cuttings or teach me to graft. On this day, there were no more promises. He lamented that he couldn't walk in his own garden lately. He had to watch it grow from his window, a spectator. He was full of regret that the cherry season was passing, that he wouldn't be able to make the thirty bottles of wine he wanted. I volunteered to buy the cherries for him when I next drove up the mountain to Shebaa, but he shook his head.

"I can't work on them," he said.

He repeated the words twice, as if apologizing.

Dr. Khairalla called a few days later. I had been worried because his car was gone from his house and his gate was closed. He told me that he was in the hospital in Nabatiyeh, and I went the next morning.

Painted in a faded white, the room was sad, and as I stepped through the door I remembered the hospital rooms of a wartime Baghdad, spare and clinical, no personal touches to ease the leavetaking of the dying. There he was, with Ivanka, who sat on a couch in a room whose only color was a band of green that circled the wall. He tried to get out of the bed to stand and greet me, but he couldn't. Frail, his skin pallid, he finally pulled himself up and wearily sat against the headboard.

"I'm exhausted," he told me.

He had a fever and a urinary tract infection. His liver was swollen, and he feared the cancer had spread to it.

Even on this day, the last I would see him, I could call him nothing but Doctor. It was fitting.

"I'll see you in September, Doctor. It's just a couple months away."

"I hope," he said.

"I'm sure I will, Doctor," though I knew otherwise. So did Dr. Khairalla.

He shrugged his shoulders. "I hope."

I caught a taxi back to Marjayoun, and the driver read the look on my face.

"Did you know someone there?" he asked.

I managed only to utter his name.

"You have to put it in the hands of God," he told me. "All the people there, no medicine will help them. No surgeries, no doctors. Your Lord will bring his health back. Only your Lord.

"A doctor," he said, "is just the *wasila*."

Wasila — an instrument, the means.

As we drove along a bend in the road, I lit a cigarette and remembered Dr. Khairalla pointing to a denuded hillside, long ago ravaged by war.

"You see here, this hill?" he had asked me.

It was once full of almond trees, he had said, their trunks showing their years, the trees planted a generation before him. In the spring they were drenched in flowers impossibly white.

The vision was otherworldly, his Jedeida.

"There are still some remnants of the trees, you see," he had told me.

EPILOGUE

In February 2011, seemingly out of nowhere, there came a time of transformation as Egypt reimagined home. It announced itself as a revolution took hold, before a tyrant was toppled, and before a lost people knew they would triumph over all the deaths and detentions, miseries and disappointments, that had characterized life in their nation. Its gathering spirit emerged most palpably in Tahrir Square in Cairo, the faded heart of a city where too many battles had been lost and too many lives humbled. What happened there was born of years of frustration, collective memories of a past some thought forgotten, and the dream of change; it was an act of imagination, of people turning acceptance into action and visions of another day. They were creating a different kind of community, linked to what once was.

Back at work a few years after completing the house in Marjayoun, I walked to the square — its name meant liberation in Arabic — not long before the revolution reached a jubilant climax. It was a few minutes after midnight on a Sunday. Rain washed the hushed streets of a place transformed. Ahmed Abdel-Moneim, draped in a blanket, felt free to talk as we crossed the Kasr el-Nil Bridge. "My vision," he told me, grinning, "goes a lot farther than what my eyes can see."

That day, cries of rebellion and, finally, of triumph had reverberated through the teeming square, a space circled by old monuments to a withering authoritarianism. "Welcome to a free Egypt," men and women had chanted. By nightfall, the scene had grown more subdued as the cheers and cries grew softer and the square became a stage for impromptu poetry readings, performances, and political debates. "What I see here," Ahmed said as we entered the square, "is what I've never seen in my life." His grin turned to a smile. "Everyone here is awake."

Canteens prepared cheese sandwiches, and no one had to pay. Volunteers ferried tea to weary guards at the barricades. Pharmacies gave out bandages and lotions, disinfectant and inhalers, intravenous solution and insulin. Artists brought their aesthetic to the asphalt, rendering work that was perhaps more inspired than memorable. As the night unfolded, vendors ambled along peaceful streets, past couples holding hands and men wearing bandages from their fights with thuggish government supporters who dared attempt to suppress their vision of this new home.

"Tea for an Egyptian pound!" one man cried. "*Koshary! Koshary! Koshary!*" shouted another, offering dishes of rice, lentils, and pasta, simple meals for hungry people. Volunteers handed out bread sticks. "My man, eat it!" shouted one. "We came for you!" There was a sense that victory was near. "Oh time, take a picture of us," went a song by Abdel-Halim Hafez, an Egyptian icon of another era, blaring from the speakers. "We will grow even closer to each other, and whoever drifts away from the square will never appear in the picture."

As I walked, I chatted with a doctor, a beautiful woman who had come home to live this moment. "They will adore this square," she told me near a line of tanks, one of them bearing the graffiti of a protester: *Egypt Is Mine*. "We'll clean the square, we'll cherish the square. It will be a symbol of making something new."

At a little past 5 A.M., as dawn's soft glow filtered across the sky, the call to prayer rang out. "Prayer is better than sleep," the muezzin nearest to me cried. Some men and women began to awaken, as the call rose across a capital known as the City of a Thousand Minarets. I walked with Mohammed Farouq toward the entrance of the bridge. Mohammed looked out at the gathering tumult, then back over his shoulder. "You feel like this is the society you want to live in," he said, gesturing to the square.

Ibn al-sa'a, goes an old Arabic phrase. The son of the hour, it means.

More figuratively, it suggests something fleeting, a lifetime captured in an instant, its fate briefly tangible. So Tahrir Square was.

Then came March, when I found myself in a town in Libya whose name I had never previously bothered to remember. Soldiers for a government crumbling but still forceful had taken me and three fellow reporters captive at a makeshift checkpoint. Bullets ricocheted around us. The soft dirt popped as they entered the earth. I had run, then stumbled on a sand berm, every muscle in my body taut. Minutes passed, and I found myself on my knees next to a simple one-room house where a woman clutched her infant child. Both cried uncontrollably. Soldiers trained their guns on us, beat us, stripped us of everything in our pockets, forced us to lie face-down. One, slighter than the others, surged toward me. "You're the translator!" he screamed. "You're the spy!" Seconds went by, but it felt far longer, and another soldier approached. Rage flared from his eyes. He shoved my face in the dirt.

"Shoot them," the soldier said calmly in Arabic.

As I lay motionless on the ground, I sensed something familiar, a feeling I recalled from Ramallah where, years before, I had lain under a cemetery-gray sky, waiting to die from the bullet wound in my back. I recalled it from Qana in 2006, where the people had cried, "Slowly, slowly!" as Lebanese soldiers, Red Cross workers, and volunteers dug with hoes, shovels, and their bare hands, searching for pieces of lost lives. I had felt it in Baghdad in 2003, when the mother of Lava Jamal, whose mauled torso was pulled from the wreckage of an American bombing, vomited at the sight of her daughter's severed head. I remembered it from Marjayoun, where I came upon a house on a hill whose grandeur had given way to insult. It was emptiness, aridity, hopelessness, the antithesis of creation, imagination.

We ended up in jail the next day, in a city called Sirte, on the Mediterranean. I suppose there are worse prisons in the Arab world. This one was relatively cheerful, painted yellow. My colleagues and I were handcuffed and left in a basement cell on ratty mattresses with a bottle to urinate in, a jug of water, and a bag of sticky dates. Tahrir Square seemed far away. Graffiti of devout prisoners were scratched into the wall. "God bring us relief," one line read in a plea to the Almighty. Scrawled next to it in tiny letters was a more intimate aside: "My beloved Firdaus."

By morning, we had been transferred to a military airport, where the beatings were worst. Blindfolded and bound with plastic handcuffs, I was hit by the butt of a gun to the head. I staggered and waited for the next blow, and the next, and wondered how many there might be. As I sat in the plane that took us to the capital, Tripoli, I panicked as the restraints dug into my wrists and numbed my swollen hands. When a man approached me, hearing my cries for help over the drone of the cargo plane's engines, I turned my head, waiting for another fist to land. I couldn't see his face, but as he leaned toward me, I could feel his breath on my ear. "I'm sorry," he whispered.

The next day, in Tripoli, shortly before Turkish diplomats negotiated our release and drove us from Libya, we sat in a lavish office as an urbane Foreign Ministry official chatted with us. His small talk suggested embarrassment, and I forgot everything he said, save a few words he quoted to my colleague in idiomatic British English.

They were two lines from a poem by William Butler Yeats: "Those that I fight I do not hate, / Those that I guard I do not love."

I hated him, though. I hated the billboards I saw as I left the country after a week in captivity, the propaganda of a regime that did not deserve to be mourned. *Forty-one Years of Permanent Joy,* read one slogan superimposed over a sunburst. *Democracy Is Popular Rule, Not Popular Expression,* read another. I hated what this had cost. I wanted to go home, and so I went to Marjayoun with my new wife and infant son. There had been no question of where we would go after my release from Libya.

When they arrived in Marjayoun, the forefathers of Isber Samara carried with them the nomadic ways of the Houran and its Bedouin residents. Their possessions were few, but each family was said to have brought the wooden *mihbaj,* to prepare their coffee, and the iron *saj,* to bake their bread. The very sound of grinding coffee was considered an invitation to anyone and everyone to come. *Stay,* it suggested. *Seek shelter.* I thought of this as I returned to Marjayoun; I thought of what was lost and what might, somehow, return. I envisioned desert wanderers of different faiths and creeds offering aid and succor to each other as they crossed the steppe. I recalled the silent respect of the women in Tyre mourning in black before eighty-six numbered coffins, destined for a single grave. I remembered Tahrir Square and what had once more, for a moment, been imagined.

As I had so often, I walked beside Isber's house of stone, passing the two most ancient olive trees, still standing from the day my grandmother had said goodbye. I thought of my daughter, soon to arrive, walking up the steps from which her great-grandmother had departed, waiting to hear Raeefa's songs. In my mind's eye I saw Laila, suddenly grown, beside these trees and repeating the Arabic words that I would one day teach her, words that would take her back to Isber's world, where the Litani River runs, over Marjayoun, over what was once our land.

This is *bayt*. This is what we imagine.

September 2011

AFTERWORD

Today, more than eight months after my husband's death, the olives on the sprawling branches of the two trees that stand at the entrance of Isber's house in Marjayoun have ripened. My son, Malik, and I will harvest them in a few days. According to a tradition renewed by Anthony when he began to harvest here, we will set a blue tarp on the ground to catch the olives that fall on their own. We will not think of beating the branches to make the olives fall. Instead, every olive left on the trees will be handpicked and dropped in a plastic bag according to its color and size.

Then we will take them to a presser in the neighboring village of Qlaya. It is the same establishment that Anthony, Malik (then only seven months old), and I used two years ago, the first time after the completion of Isber's house that we harvested the olive trees there as a family. Three friends, from Beirut, Baghdad, and Cairo, joined us in Marjayoun for the harvest that year. Another friend, from Jerusalem, was scheduled to come but missed his flight after being delayed at an Israeli checkpoint. Anthony believed in turning the harvest into a ritual completed by a community of family and friends. He spoke, last October, about organizing another trip for this year's harvest. (Isber Samara's trees, like most here, produce fruit every other year.)

There was something about picking the olives, Anthony always said, that made him feel a connection to the past, witnessed by the house where the trees had so long stood. Sometimes, particularly during the writing of *House of Stone,* it seemed that Anthony imagined this past so vividly that it became more real to him than the present. Isber was, in a sense, Anthony's guide, his way into an era, revealing the values that Anthony saw as essential if the Middle East is ever again to see peace. They are the same values that Isber built into the walls of his house: hospitality, cosmopolitanism, and tolerance.

Anthony often joked that he wished he had been born an Ottoman gentleman. Actually, I believe he was. As a human being, he carried—naturally, gracefully—the humane and welcoming spirit that defined the Middle East of the past. This was Isber's world; as Anthony labored on this book, which meant so much to him, that world became his dwelling place as well. I like to think that the place where Anthony finds himself today, wherever that may be, is one with the beauties of the world that Isber ultimately lost.

Isber Samara imagined a future for the house and for himself. At first he saw the place he created as a testimony to his own success. As Anthony wrote, Isber spent his life "trying to make his own name and achieve wealth and the sort of success admired by others." But Isber, like my husband, saw the world change and was himself changed by the transformation of his homeland. The Ottoman gentlemen he so admired soon lost their stature, and what he had worked so hard to achieve all his life suddenly lost its importance. As his family was broken apart by fear, lawlessness, and conflict ravaging the countryside near Marjayoun, he became a different man, a family man who valued the future of his children more than his own need to surround himself with them. Three of the young Samaras left for America. He would never see them again. And in 1928 he passed away from a pneumonia that was blamed on the wind. He could not have lived in the house for more than a few years. Like Anthony, his stay in the House of Stone was short-lived.

Of the many similarities between Anthony and his great-grandfather, their ambition is the most obvious. But what is striking to me now is that they both witnessed the most terrible violence the world can bestow, but from it both created lasting, positive statements. Isber left his house and his children to change the world. Anthony not only left these legacies; he also left his work, writing that will stand the test of time and will forever guide us all toward peace as it reminds us of the human costs of war and conflict.

I often teased Anthony, telling him that his love for the house was greater than his love for anyone and anything else. He would laugh, a little amused by the thought. I think a part of him felt that the house deserved that kind of love and commitment. It was where he found his family, and it was also where he found himself, where he realized his most meaningful ambition: the hope of becoming a good man who, like his great-grandfather, could value his family's happiness and future far more than his own. He found a part of himself in that house, a part that perhaps he had thought was lost forever amid the wars and bloodshed he had covered for more than a decade as a foreign correspondent. The house became his home before it was even livable. He lived there long before the walls rose and the furniture arrived. I like to think he lives there still.

I have been asked repeatedly about the fate of the house since Anthony passed away last February, during his final reporting trip to Syria for the *New York Times,* to write about the armed opposition that was setting up a stronghold in the north of that country. I have no doubt that a house that survived such mayhem in the past century will go on to last for generations to come. On our first visit to Marjayoun without Anthony, four months after my gentle husband passed away, Malik stood in the garden of the house for a while, calling out, "Daddy, where are you? Daddy, where are you?" I realized at that moment that no matter how hard I reassured myself that this place was still our home, it was simply not true. It was not home anymore. It had lost its soul when it lost Anthony. It never will be the same place again. But someday Malik may find his father there, just as Anthony found Isber and learned, from his house, from his life, how to be a man.

I suppose the loneliness that fell on the house when Isber Samara died is the same loneliness that haunts it now. Sometimes it reminds me of what a bride left at the altar might look like years later: still in her wedding gown, no longer white, her hair undone, her face distressed.

Anthony said that something evil touched his soul during the week he spent in captivity in Libya, in March of 2011, along with three other colleagues from the *New York Times*. When I later asked him what he meant by that, he said he felt that death had touched him and that the feeling lingered with him.

In the few months that preceded his death, he often said he was overwhelmed with a feeling of dread. As long as he lay in bed in the morning, he could mute that premonition, but once he got up, it entirely consumed him. And so almost every day we'd go over a list of things—events, places, and people—trying to pinpoint what was worrying him, without any success. But he went on. Despite his fears, he went on to tell stories, to tell the truth of places where people were dying without mercy, where humanity was fading.

Anthony writes that his great-grandfather "must have longed for and been haunted by what he had barely touched but not had the chance to savor." I sometimes wonder how much Anthony was writing about himself when he was writing about Isber Samara. Certainly, like his great-grandfather, he wanted to leave something behind. In his case, it was his house and his book.

Anthony often said to me that *House of Stone* was going to be his chef-d'oeuvre. Perhaps he knew, at some level, that the book would be the last thing he was going to publish. He worked so tirelessly on it because it was about and for his family, and it is to his family that he leaves it.

During our last visit to Marjayoun, last February, just a week before he left for Syria, Anthony, with Malik at his side, spent most of his time in the garden, pulling weeds and pruning plants. He watered the turnips he had planted a few months before, and said that when he returned from Syria they would be ready to be picked and pickled. And he did make the best pickled turnips. When I left the house to check on them, Malik's mouth was covered with dirt. Apparently he had eaten some, and the sight amused Anthony. On our way back to Beirut, where we lived and worked, I asked Anthony not to let Malik eat dirt again. But then it occurred to me that Malik, from that point on, would be forever connected to the house because he had eaten from the earth where it stands. I shared this notion with Anthony, and after a long silence he said, "I don't know why I haven't thought about this before. I should eat Marjayoun dirt too. Next time."

Of course he never went back, at least not to eat dirt from the house or to pick the turnips. He went back to lie in his great-grandfather's ground, under the olive trees that he loved and which we will pick from soon.

We will leave the fruit on the branches, which cast shade on Anthony's grave, marked with twelve cemento tiles of different colors and patterns. I like to think that the olives will gently fall to the ground next to him, and that somehow my dear husband will savor them.

Nada Bakri
October 2012

NOTE TO READERS

This book began as a passing thought in 2006, and more than anyone else, two men deserve the credit for its finally becoming a book all these years later. They are George Hodgman and Robert Shepard, two of the finest men I know. As my agent, Robert showed faith that might have been misplaced, and friendship that I might not have deserved. George is simply the most brilliant editor alive today. He despaired, cajoled, harangued, and inspired until the book went from an unwieldy mess into something else. As with *Night Draws Near,* his name should grace the cover.

My daughter Laila was my inspiration to begin this book. The love and commitment of my wife Nada allowed me to finish it. As we did, our son Malik was there, seemingly aware that his family's story was being told. I learned from my father Buddy and my mother Rhonda to take pride in the past; they represent everything that family is supposed to mean. Their spouses, Shara Shadid and Charles Moschera, have always treated me, my sister Shannon, and my brother Damon as their own children.

I always meant *House of Stone* as a testament to our sprawling clan in Oklahoma City, which will always stay together even when it is apart. Indeed, the spirit of so many who have passed away remains with us — my grandparents, Albert, Raeefa, George, and Onie Dee, and my aunts and uncles, especially Nabeeh, Nabeeha, Najiba, and Nannette. Those still living spent endless hours with me, and I am especially grateful to Aunt Gladys, Aunt Adeeba, Aunt Elva, Uncle Charlie, Uncle Edward, Aunt Yamama, Aunt Georgeann, and Ghassan Mike Samara. Of everyone in my family, the house means the most to my cousin Joumana Lahoud. It would have never been rebuilt without her, her husband Fouad, and Fouad's brother, Armando. It will always be there for my cousins, who are like brothers and sisters to me. Nour Malas spent long hours immersed in the project, with invaluable help in the research. Michael Provence, Leila Fawaz, and Carol Hakim were remarkably generous in reviewing the book's historical sections. Finally, I have to thank the family that I have come to know in Jedeidet Marjayoun, the people who helped me understand the meaning of imagination and the communities it can create. I count among my own all the friends I came to know there, the *maalimeen* with whom I worked and, of course, Abu Jean and Dr. Khairalla.

As I mentioned in my earlier books, transliterating Arabic into English is typically a messy business. This book makes it even more so. My wife and I sought to render the words as they sounded colloquially, faithful to the pronunciation, but we never agreed on any real standard. For speakers of Arabic, we trust the words will be familiar, at least phonetically. I have rendered the same name differently to distinguish characters — Nabeeh and Nabih, or Nabeeha and Nabiha, for instance. Though most of the characters in the book are identified by their real names, I occasionally relied on pseudonyms for others, hoping not to make anything awkward for them in a town that is quite small.

Stories passed down from generation to generation represented the bulk of the research for *House of Stone*. But many of the sections would have been impossible without the impressive scholarly research into the region and its history. I am especially indebted to Hanna Hardan Khoury and his exhaustive work *The Rich News of the Families of Marjayoun and Wadi al-Taym*, as well as Henri Abou Arraj and his remarkable collection *Old Marjayouni Papers*. I relied extensively on the following books: Michael Shadid, *Crusading Doctor*; Farid Hourani, *Olinda's Dream*; Cecil Hourani, *An Unfinished Odyssey*; Farid Haddad, *At the Foot of Mount Hermon*; Friedrich Ragette, *Architecture in Leba-*

non; Herb Ham, *Worshipping the Undivided Trinity;* Tom Caldwell, *From the Hills of Lebanon;* Michael Provence, *The Great Syrian Revolt and the Rise of Arab Nationalism;* Elizabeth Thompson, *Colonial Citizens;* Philip S. Khoury, *Syria and the French Mandate;* and Beate Hamizrachi, *The Emergence of the South Lebanon Security Belt.* For many of the historical sections I owe a large debt to the unpublished work of Tom Caldwell, Jabour Shadid, and Raymond Habiby.